Radiotelemetry Applications for Wildlife Toxicology Field Studies

Recent titles from the Society of Environmental Toxicology and Chemistry (SETAC)

Principles and Processes for Evaluating Endocrine Disruption in Wildlife
Kendall, Dickerson, Giesy, Suk, editors
1998

Quantitative-Structure Activity Relationships in Environmental Science-VII
Chen and Schüürmann, editors
1997

Atmospheric Deposition of Contaminants to the Great Lakes and Coastal Waters
Baker, editor
1997

Chemically Induced Alterations in Functional Development and Reproduction of Fishes
Rolland, Gilbertson, Peterson, editors
1997

Chemical Ranking and Scoring: Guidelines for Relative Assessment of Chemicals
Swanson and Socha, editors
1997

Ecological Risk Assessment of Contaminated Sediments
Ingersoll, Dillon, Biddinger, editors
1997

Public Policy Applications of Life-Cycle Assessment
Allen, Consoli, Davis, Fava, Warren, editors
1997

Reassessment of Metals Criteria for Aquatic Life Protection: Priorities for Research and Implementation
Bergman and Dorward-King, editors
1997

Multi-Media Fate Model: A Vital Tool for Predicting the Fate of Chemicals
Cowan, Mackay, Feijtel, van de Meent, Di Guardo, Davies, Mackay, editors
1995

For information about additional titles or about SETAC's international journal,
Environmental Toxicology and Chemistry,
contact the SETAC Office, 1010 N. 12th Avenue, Pensacola, FL USA 32501-3367
T 850 469 1500   F 850 469 9778   E setac@setac.org   http://www.setac.org

Radiotelemetry Applications for Wildlife Toxicology Field Studies

Edited by

Larry W. Brewer
Ecotoxicology & Biosystems Associates, Inc.
Kathleen A. Fagerstone
National Wildlife Research Center

Proceedings of the Pellston Workshop on Avian Radiotelemetry in Support
of Pesticide Field Studies
5–8 January 1993
Pacific Grove, CA

SETAC Special Publications Series

SETAC Liaison
Kathleen A. Fagerstone
National Wildlife Research Center

Current Coordinating Editor of SETAC Books
Christopher G. Ingersoll
U.S. Geological Survey, Midwest Science Center

Published by the Society of Environmental Toxicology and Chemistry (SETAC)

Cover design by Mike Kenney
Indexing by IRIS

Library of Congress Cataloging-in-Publication Data

Pellston Workshop on Avian Radiotelemetry in Support of Pesticide Field Studies (1993 : Pacific Grove, Calif.)
Radiotelemetry applications for wildlife toxicology field studies : proceedings from a Pellston Workshop on Avian
Radiotelemetry in Support of Pesticide Field Studies, 5–8 January 1993. Pacific Grove CA/ edited by Larry W.
Brewer, Kathleen A. Fagerstone.
p. cm. -- (SETAC special publications series)
"Published by the Society of Environmental Toxicology and Chemistry (SETAC) and the SETAC Foundation for
Environmental Education."
Includes bibliographical references and index
ISBN 1-880611-20-1
1. Birds--Effect of pesticides on--United States--Congresses. 2. Animal radiotracking--United States--Congresses. 3.
Pesticides and wildlife--United States--Congresses. I. Brewer, Larry W. II. Fagerstone, Kathleen A., 1951– . III. SETAC
(Society) IV. Title. V. Series.
QL682.P45 1993
598.17--dc21   98-4522
                                                                                                                  CIP

© 1998 Society of Environmental Toxicology and Chemistry (SETAC)
SETAC Press is an imprint of the Society of Environmental Toxicology and Chemistry.
No claim is made to original U.S. Government works.

International Standard Book Number 1-880611-11-2
Printed in the United States of America
05 04 03 02 01 00 99 98   10 9 8 7 6 5 4 3 2 1

♾ The paper used in this publication meets the minimum requirements of the American National Standard for
Information Sciences—Permanence of Paper for Printed Library Materials, ANSI Z39.48-1984.

Reference listing: Brewer LW, Fagerstone KA. 1998. Radiotelemetry applications for wildlife toxicology field studies.
Proceedings from a Pellston workshop on avian radiotelemetry in support of pesticide field studies; 1993 January
5–8; Pacific Grove CA. Pensacola FL: Society of Environmental Toxicology and Chemistry. 224 p.

The SETAC Special Publications Series

The SETAC Special Publications Series was established by the Society of Environmental Toxicology and Chemistry (SETAC) to provide in-depth reviews and critical appraisals on scientific subjects relevant to understanding the impact of chemicals and technology on the environment. The series consists of single- and multiple-authored or edited books on topics reviewed and recommended by the SETAC Board of Directors for their importance, timeliness, and contribution to multidisciplinary approaches to solving environmental problems. The diversity and breadth of subjects covered in the series reflect the wide range of disciplines encompassed by environmental toxicology, environmental chemistry, and hazard and risk assessment. Despite this diversity, the goals of these volumes are similar; they are to present the reader with authoritative coverage of the literature, as well as paradigms, methodologies and controversies, research needs, and new developments specific to the featured topics. All books in the series are peer reviewed for SETAC by acknowledged experts.

The SETAC Special Publications are useful to environmental scientists in research, research management, chemical manufacturing, regulation, and education, as well as to students considering careers in these areas. The series provides information for keeping abreast of recent developments in familiar subject areas and for rapid introduction to principles and approaches in new subject areas.

SETAC would like to recognize the past SETAC Special Publications Series editors:
T.W. La Point
    The Institute for Environmental and Human Health
    Texas Tech University
    Lubbock, TX

B.T. Walton
    U.S. Environmental Protection Agency
    Research Triangle Park, NC

C.H. Ward
    Department of Environmental Sciences and Engineering
    Rice University
    Houston, TX

Contents

List of figures

List of tables

Foreword

On 6-8 January 1993, the Pellston Workshop, Avian Telemetry in Support of Pesticide Field Studies, was held at the Asilomar Conference Center, Pacific Grove, California. The workshop was an outgrowth of the Avian Effects Dialogue Group, a group of personnel from government, industry, academia, and environmental organizations convened by Resolve, an independent program of the World Wildlife Fund, to discuss bird/pesticide issues. The workshop was organized by 8 members of the Telemetry Subcommittee of the Avian Effects Dialogue Group and cochaired by David Fischer (Miles/Mobay) and Bill Williams (ecological planning & toxicology, inc.). Contributors included the Society of Environmental Toxicology and Chemistry (SETAC), the Avian Effects Dialogue Group, Ciba Geigy Corporation, DuPont Agricultural Products, ICI Agrochemicals, Miles Incorporated, Rhône Poulenc Ag Company, Rohm & Haas Company, the U. S. Department of Agriculture, and the U. S. Environmental Protection Agency (USEPA).

Participants evaluated options and developed recommendations for the use of radiotelemetry techniques in field studies of pesticide effects on birds. Terrestrial field studies of pesticides are a relatively new requirement of USEPA, the guidelines having been published in 1988. Data developed from carcass searches, the main technique used for evaluating effects on birds, are difficult to interpret; yet standard approaches to the use of telemetry were not included in the USEPA guidelines. Workshop organizers had anticipated that radiotelemetry would provide a more definitive approach to field investigations of pesticide effects. In the past, the lack of simple, reliable radio systems small enough to use on birds forced investigators to design and construct radiotelemetry systems built for the birds' special needs. Many innovative approaches for use of telemetry systems for bird studies resulted, but the approaches also minimized the ability of the systems to be used by other investigators. In fact, the effort and expense associated with building unique systems sometimes resulted in reluctance to share designs with other groups. One focus of this workshop was to share innovations with those who design and use radiotelemetry for bird studies and to exchange ideas about how new technologies might be used in the future. The workshop brought together about 50 toxicologists, ecologists, representatives of industry and regulatory agencies, telemetry engineers, and statisticians to discuss how telemetry technology best could be used in field studies to delineate exposure of birds to chemicals and subsequent effects.

The workshop consisted of 2 days of invited participants who presented papers, followed by a day of group discussions of case studies and development of general recommendations for radiotelemetry use in pesticide field studies. The amount of information disseminated at this workshop was impressive. Numerous uses of telemetry were discussed, ranging from novel uses of currently available telemetry to ideas about technological developments that would provide better information for field studies. The participants agreed that the use of telemetry in field studies can provide information about both exposure (documenting visits to treated areas) and effects (tracking survivability and mortality). Additional information about physiological condition and behaviors can also be gained from properly designed telemetry systems. Use of motion switches and/or mortality switches is helpful for determining the status of birds exposed to contaminants. Automated collection of activity information may provide information about bird well-being (low-activity levels may be the result of lethargy or sickness). The workshop members developed a list of ideas that might help direct telemetry development. Some of the general recommendations are listed below:

1) The Global Positioning Satellite System (GPS) appears to be a very promising new technology for providing information about animal location. This technology would integrate well with implanted or attached data storage chips to provide an accounting of sites visited by tracked birds. The ability to accurately position an individual bird and send this signal to a computer continuously could revolutionize field studies. However, further reduction in transmitter size will have to be achieved to attain these advancements in bird studies.

2) Activity sensors within the transmitter that provide a signal proportional to relative activity and movement could be used to monitor behavior.

3) Information about bird health and/or position might be contained in data storage chips within the transmitter that could be queried at intervals.

4) Implanted transmitters used to collect physiological information would cause less interference with normal behaviors, provide less external drag, and would not catch on vegetation. The transmitting antenna could be externalized to transmit greater distances.

5) Nest switches or nest-attentiveness monitors could provide information about time spent at nest sites by adults. Audio sensors at nests could pick up and send hatchling calls and other data to a receiving station.

6) Signal strength information could potentially provide information about relative distance of the transmitter from the antennae, as a supplement to plots of bird location.

7) The use of smart duty cycles to transmit signals only when trackers are likely to be interrogating the birds and to transmit at fixed, short intervals would provide considerable savings in transmission power and battery life.

8) Use of remote dosimeters could be a viable approach to determining the exposure history of transmittered animals.

This book contains peer-reviewed papers by the authors of many of the lectures presented at the meeting. The authors offer a wide variety of expertise and experience, ranging from regulatory to field, and a broad array of viewpoints. What all authors stress is that radiotelemetry offers one of the best techniques available for assessing avian exposure and survival after a pesticide treatment. Radiotelemetry improves the chances for finding avian carcasses if mortality occurs and provides one of the few techniques for measuring avian exposure. Our hope for this book is that it will be helpful to the field biologist conducting regulatory studies by 1) identifying the information of importance to regulatory agencies, 2) suggesting improvements for study design, 3) providing information on equipment and methods, and 4) suggesting data analysis techniques.

We are indebted to the chapter authors for their persistence and patience during this process. We thank Ray Sterner and Larry Kolz of the National Wildlife Research Center (NWRC) for editing the book, and Marilyn Harris, also of the NWRC, for a superb job of formatting the initial drafts of chapters. Chris Ingersoll acted as liaison for SETAC during the development of the book. Finally, we thank the SETAC staff, including Linda Longsworth, Chris Englert, and Stacey Hagman for their assistance with the workshop and this publication.

Kathleen Fagerstone and Larry Brewer

Preface

Kenward (1987) described 2 uses and 2 definitions for animal radio transmitters (radio tags). The uses are 1) locating study animals in the field, or "radio tracking", and 2) monitoring physiological or behavioral characteristics of animals, or "radiotelemetry." Radio-tracking techniques have been employed regularly in wildlife research for the past 3 decades. Advances in technology, particularly micro-electronics, have allowed wildlife biologists to monitor free-ranging animals during the course of their normal activities and thus have provided significant advances in our understanding of species specific behavior, physiology, and ecology.

In the 1980s, the U. S. Environmental Protection Agency (USEPA) was charged with the responsibility of determining that a pesticide does not cause "unreasonable adverse effects on man or the environment" (Federal Insecticide, Fungicide and Rodenticide Act [FIFRA] 1988). Registrants who requested registration or reregistration of pesticide products had to provide data upon which USEPA could base a determination regarding "unreasonable and adverse effects" to wildlife. Toward the end of the 1980s, USEPA was requesting that wildlife field studies be conducted by potential registrants to produce these data. An expanded and updated guideline for such field studies was provided by USEPA in 1988 (Fite et al. 1988). The contents of this book include the description and results of several radio-tracking and radiotelemetry studies designed around these guidelines.

In 1992, there were policy changes in USEPA regarding the need for field studies in the pesticide registration decision-making process (see Chapter 1). At this time USEPA decided to make decisions based on laboratory data and stated that actual field data rarely would be needed. This new policy strongly discouraged further applications of radio tracking or radiotelemetry in support of pesticide registration. However, the USEPA Scientific Advisory Panel (SAP), in its 1997 meeting made the following comments and recommendations to USEPA (Bailey et al. 1997):

1) The Office of Pesticide Programs should reinstate field studies.
2) The practice of extrapolating environmental risk factors from laboratory data has not been validated.
3) The Office of Pesticide Programs should develop the necessary databases and methodologies to conduct probabilistic assessments of risk relative to pesticide registration.

Much of the data necessary to measure or model pesticide exposure and survival of wild birds can be produced best via radiotelemetry and radio-tracking studies. Therefore, these techniques may become an important aspect of the pesticide registration process. The studies and discussions presented in this book provide a foundation of methodology on which to build even more refined techniques for evaluating the potential risk of pesticides to wildlife. Part 1 provides an overview of avian pesticide studies and wildlife radiotelemetry in general. Part 2 contains examples of field studies employing radiotelemetry. In this section, Chapters 5, 6, and 8 are field studies conducted under FIFRA Good Laboratory Practices (GLP) Guidelines, while Chapters 4, 7 and 9 are not GLP guideline studies. Part 3 offers insight into statistical design of radiotelemetry field studies, and Part 4 covers engineering, data handling, and emerging technology aspects of radiotelemetry.

Future development needs for highly definitive and quantitative avian toxicology field studies include fully automated telemetry/tracking systems and further miniaturization of transmitters. When these technological advancements are available, our ability to produce precise estimates of individual animal exposure and response to pesticide applications will be strongly enhanced.

References

Bailey T, Fite E, Mastradone P, Montague B, Parker R, Sunzenauer I, Wolf J. 1997. EFED's summary of the SAP review of ecological risk assessment methodologies. Washington DC: USEPA. 9 p.

[FIFRA] Federal Insecticide, Fungicide and Rodenticide Act. 1988. Public Law 92-516, Title 7, United States Code. Washington DC: USEPA, Office of Pesticide Programs, PMSD, Information Services Branch.

Fite EC, Turner LW, Cook NJ, Stunkard C, Lee RM. 1988. Guidance document for conducting terrestrial field studies. Ecological Effects Branch Hazard Evaluation Division, Office of Pesticide Programs. Washington DC: USEPA. 134 p.

Kenward R. 1988. Wildlife radio tagging. New York: Academic. 222 p.

Acknowledgments

The workshop and this publication were made possible by financial support from the following organizations:

CIBA-GEIGY
DuPont Agricultural Products
ICI Agrochemicals
Miles, Inc.
Rhône-Poulenc
Rohm & Haas
U.S. Department of Agriculture
U.S. Environmental Protection Agency

Editors

Larry Brewer is President and Senior scientist at Ecotoxicology and Biosystems Associates, Inc. He received his Bachelor of Science degree from Iowa State University and his Master of Science from the University of Washington in Wildlife Science. He was employed as a research biologist by the Washington Department of Wildlife from 1972 to 1986. Mr. Brewer served as Field Research Leader at the Institute of Wildlife Toxicology, Western Washington University from 1986 to 1989, and in 1989 he joined the faculty of Clemson University where, through 1992, he served as Ecotoxicology Research Section Leader for The Institute of Wildlife and Environmental Toxicology. Mr. Brewer is a member of the Society of Environmental Toxicology and Chemistry (SETAC), The Wildlife Society (TWS), and the American Birding Association (ABA). He has served on the Editorial Board for SETAC and participated in several SETAC Pelston Workshops. Mr. Brewer was charter chair of The Wildlife Toxicology Working Group for TWS, serving in that capacity for 3 years. His professional interests include wildlife ecotoxicology, radio-telemetry in wildlife research, ecological risk assessment relative to contaminants, and species-specific ecology.

Kathleen Fagerstone is Manager of the product Development Research Program at the National Wildlife Research Center in Fort Collins, Colorado. She obtained her Bachelor of Science in Zoology from Colorado State University, Fort Collins, and her Master of Science and Ph. D. from the University of Colorado, Boulder. In her current position, Dr. Fagerstone oversees development of new technology for solving wildlife damage problems, including repellent registrations, drug authorizations, and immunocontraceptive vaccines. Her past research interests have been in the areas of risk assessment and small mammal ecology, with research on prairie dogs, ground squirrels, and the black-footed ferret. She is past president of the Colorado Chapter of The Wildlife Society and has served as an Associate Editor for the *Journal of Wildlife Management.*

Participants and Contributing Authors*

Charles Amlaner
University of Arkansas
Fayetteville AR

Fred Anderka
Holohil Systems Ltd
Woodlawn, Ontario
Canada

Richard Balcomb
Ciba-Geigy Corporation
Greensboro NC

Richard Bennett
USEPA
Corvallis OR

Louis Best
Iowa State University
Ames IA

Lawrence Blus
US Fish and Wildlife Service
Corvallis OR

Larry Brewer
EBA, Inc.
Sisters OR

Christine Bunck
US Fish and Wildlife Service
Laurel MD

Mike Burke
Wildlife Materials, Inc.
Carbondale IL

Crystal Driver
Battelle Norwest
Richland WA

Joseph Dulka
DuPont Agricultural Products
Wilmington DE

Peter Edwards
Zeneca Agrochemical
Bracknell Berks
United Kingdom

Kathleen Fagerstone
US Dept of Agriculture
Denver CO

Anne Fairbrother
USEPA
Corvallis OR

Dave Farrar
USEPA
Washington DC

David Fishcer
Miles, Inc.
Stilwell KS

Ed Fite
USEPA
Washington DC

Michael Fry
University of California
Davis CA

Mark Fuller
Bureau of Land Management
Boise ID

James Gagne
American Cyanamid Company
Princeton NJ

Lisa Ganio
Mantech Environmental Technology, Inc.
Corvallis OR

David Gilmer
US Fish and Wildlife Service
Dixon CA

Peter Greig-Smith
Ministry of Agriculture Fisheries & Food
Lowestoft, Suffolk
United Kingdom

Christian Grue
University of Washington
Seattle WA

Ronald Kendall
Clemson University
Pendleton SC

Robert Kenward
Institute of Terrestrial Ecology
Wareham, Dorset
United Kingdom

Eph Konigsberg
Konigsberg Instruments, Inc.
Pasadena CA

Hank Krueger
Wildlife International, Ltd.
Easton MD

Larry Kuechle
Advanced Telemetry Systems, Inc.
Isanti MN

James Lotimer
Lotek Engineering, Inc.
Newmarket, Ontario
Canada

Lyman McDonald
WEST, Inc.
Cheyenne WY

Ellen Mihaich
Rhône-Poulenc
Research Triangle Park NC

Pierre Mineau
Canadian Wildlife Service
Ottawa, Ontario
Canada

Raymond O'Connor
University of Maine
Orono ME

Rodney Parrish
SETAC/SETAC Foundation
Pensacola FL

Richard Poché
Genesis Laboratories
Ft. Collins CO

Edward Schafer, Jr.
US Dept of Agriculture
Denver CO

John Skatski
University of Washington
Seattle WA

Bruce Stanley
DuPont Agricultural Products
Wilmington DE

Paul Toll
Miles, Inc.
Stilwell KS

Doug Urban
USEPA
Washington DC

Michale Whitten
KBN Engineering & Applied Sciences
Gainesville FL

Bill Williams
ecological planning & toxicology
Corvallis OR

David Wustner
VALENT USA Corporation
Walnut Creek CA

* Please note that these are the affiliations of the participants and contributing authors at the time of the workshop.

Abbreviations

AChE	acetylcholinesterase
ADF	automatic direction finding
AHY	after hatching year
ANODEV	analysis of deviance
ANOVA	analysis of variance
antiChE	anticholinesterase
APS	automatic positioning system
ARRL	American Radio Relay League
ASCS	Agricultural Stabilization Conservation Service
ATV	all terrain vehicle
BChE	butrylcholinesterase
C/A	coarse acquisition
CDFA	California Department of Food and Agriculture
ChE	cholinesterase
CRP	Conservation Reserve Program
Db	decibel
Dbm	decibel relative to 1 milliwatt
DSP	digital signal processing
DWRC	Denver Wildlife Research Center
EEC	estimated environmental concentration
EIRP	effective isotropic radiated power
ERA	ecological risk assessment
ERP	effective radiated power
FAA	Federal Aviation Administration
FIFRA	Federal Insecticide, Fungicide and Rodenticide Act
GIS	geographic information system
GLP	Good Laboratory Practices
GPS	global positioning system or global positioning satellite
HMAA	harmonic-mean activity area
HY	hatching year
LEO	low-earth orbit

LOC	level of concern
MCP	minimum convex polygon
MDS	minimum detectable or discernable signal level
MHz	megahertz
MSA	mechanically scanned array
NOEL	no-observed-effects level
OC	organochlorine
OP	organophosphate or organophosphorus
OPP	(USEPA's) Office of Pesticide Programs
ORD	(USEPA's) Office of Research and Development
PTT	platform transmitter terminal
RED	re-registration eligibility decision
RF	radio frequency
S/A	selective availability
SAP	(USEPA's) Scientific Advisory Panel
SAS	statistical analysis system
SD	standard deviation
SNR	signal power to noise power ratio
SURPH	Survival by Proportional Hazards
SwRI	Southwest Research Institute
SYA	switched yagi array
USEPA	U.S. Environmental Protection Agency
USFS	U.S. Forest Service
USFWS	U.S. Fish and Wildlife Service
USGS	U.S. Geological Survey
UTM	universal transverse mercator
VSWR	voltage standing wave radio

Chapter 1

Radiotelemetry in avian field studies to support pesticide registration decisions: a regulatory perspective

Douglas J. Urban

Monitoring avian species is not new with respect to data requirements for pesticide field studies and public guidance. However, such field studies are often viewed as problematic. These studies have provided useful confirmatory data for making conclusions concerning acute risk to birds from individual pesticides. The U.S. Environmental Protection Agency (USEPA) has required these studies when a pesticide exceeds the regulatory levels of concern (LOC) for birds. With recent policy changes within USEPA's Office of Pesticide Programs (OPP), situations where field studies are required and the questions being asked for making regulatory decisions are changing. Office of Pesticide Programs will now rely on laboratory data, existing field-study data, bird-kill monitoring data, risk mitigation measures, risk-characterization factors, and post-registration monitoring for making these decisions. This paper will present a view of the likely changes in the ecological risk assessment (ERA) paradigm, and the role of avian telemetry in improved risk assessments and regulatory decisions.

The stated purpose of the Pellston workshop on avian radiotelemetry in support of pesticide field studies is to "evaluate options and develop recommendations for the use of radiotelemetry techniques in field studies of pesticide effects on birds." Among other more obvious factors, success depends upon a cultivated understanding of each participant's viewpoint. This paper provides a vista from the office-bound station of the regulatory scientist. With the ecologist, the toxicologist, and the telemetry specialist providing their perspectives, only the registrants are officially missing.

From the viewpoint of a wildlife ecologist in the USEPA's Office of Pesticide Programs (OPP), avian pesticide field studies have provided useful confirmatory data for making avian acute risk conclusions for individual pesticides (see Figure 1-1). Most avian pesticide field studies have been conducted to refute the presumption that acute risks to birds will occur under actual use conditions of the pesticide. A risk quotient[1] is calculated us-

---

[1]A risk quotient is the ratio of the estimated environmental concentration of a pesticide to a toxicity test effect level that is usually acute or chronic toxicity as determined in laboratory testing for a given species. It is calculated by dividing an appropriate exposure estimate (e.g., estimated environmental concentration) by an appropriate toxicity test effect level (e.g., LC50).

ing estimated environmental concentrations (EEC) or doses and laboratory toxicity data. The presumption of risk is based upon a regulatory level of concern[2]. The risk quotient is compared to the regulatory level of concern to determine whether the risk exceeds the level of concern and thus presumption of risk.

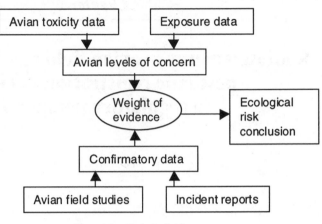

Figure 1-1   Ecological risk assessment

Background and history

The Federal Insecticide, Fungicide, and Rodenticide Act (FIFRA) assigns clear responsibilities to USEPA and the regulated parties (FIFRA 1988). For a product to be registered or reregistered, USEPA must determine that it will *not* (emphasis added) cause "unreasonable adverse effects on man or the environment." Those requesting registration or reregistration must provide data for USEPA to make a determination of "unreasonable adverse effects" (USEPA 1984). The law also requires USEPA to "publish guidelines specifying the kinds of information which will be required to support the registration of a pesticide and shall revise such guidelines from time to time" (FIFRA, P.L. 92–516, Section 3 (c)(2)(A)).

Guidelines and technical guidance documents have been published by the USEPA's OPP for terrestrial wildlife field studies, as shown in Table 1-1. Guidance for full-scale field studies was provided as early as 1975 (USEPA 1975). However, USEPA required and received far more simulated avian pen studies than full-scale field studies.

In 1982, the avian small-pen study was eliminated from the guidelines because of the many design-related limitations of the study. Further, no recommendations for improvement were submitted to the Agency (USEPA 1982). Gradually, USEPA began to request more full-scale avian pesticide field studies. Specific guidance for these studies was updated and expanded in 1988 (Fite et al. 1988). The updated guidance described two types of field studies for wildlife: a screening study and a definitive study. The screening study is essentially an effect/no-effect study designed to show that the risk to the portion of the wildlife population exposed to the pesticide is minimal under actual use conditions. In

---

[2]The LOCs are criteria used to indicate potential risk to nontarget organisms and the need to consider regulatory action. The criteria indicate that a pesticide, when used as directed, has the potential to cause undesirable effects on nontarget organisms.

Table 1-1   *Pesticide guidance documents for published terrestrial field studies*

Year	Description
1975	USEPA. 1975 (June 25). 40 CFR No. 123. Published guidelines for registering pesticides in United States.
1978	USEPA. 1978 (July 10). 40 CFR No. 132. Published registration of pesticides in the United States: proposed guidelines.
1982	USEPA. 1982. Pesticide assessment guidelines, subdivision E, hazard evaluation: wildlife and aquatic organisms. Washington DC: USEPA. EPA 540/09–82–024.
1988	Fite EC, Turner LW, Cook NJ, Stunkard C. 1988 (September). Guidance document for conducting terrestrial field studies. Washington DC: USEPA, Office of Pesticide Programs. EPA 540/09–88–109.

contrast, the definitive study is designed to quantify the magnitude of the effects identified in the screening study. Definitive studies examine a sample of the entire local population in the treated area. The 2 studies are often referred to as Level 1 and Level 2 studies, respectively. Recent policy changes concerning the need for field studies in decision-making will be discussed later (Fisher 1992).

When field studies are required

The USEPA has required avian pesticide field studies when the regulatory avian LOCs based upon laboratory data are exceeded (see Table 1-2). Until 1990, avian pesticide field studies were required to address risk concerns when the EEC or dose exceeded the LC50 or LD50/bird, respectively. USEPA considered such pesticide uses as high risk. However, for many highly toxic organophosphate (OP) and carbamate insecticides, the exposure levels often greatly exceeded the acute toxicity values. Thus, there was little urgency to specify the avian-acute-high-risk LOC.

Table 1-2   *Avian levels of concern*

Endpoint	Avian LOC	Conclusion	
Acute Effects EEC/LC50 or LD50/ft²	0.5	High avian risk	May need regulatory action
Chronic Effects EEC/NOEL	1.0	High avian risk	May need regulatory action

In 1992, USEPA announced the selection of 1 LD50/ft^2 as the cutoff LOC. This was based upon field study data submitted to USEPA at that time that indicated that pesticide applications resulting in environmental concentrations of at least 1 LD50/ft^2 have resulted in avian mortality (USEPA 1992a). Since that time, USEPA changed the LOC to 0.5 LD50/ft^2 primarily to add a level of safety to the risk estimate (Fisher 1992). This proposal has generated considerable discussion both within and outside the USEPA. A final decision awaits internal and external evaluation of this change. Until that time, the USEPA is continuing to use 0.5 LD50/ft^2, as the LOC.

Current situation

To date, USEPA has requested and received approximately 40 terrestrial wildlife pesticide field studies. About half of these studies have been reviewed. Most field studies have investigated the acute effects of OP and carbamate insecticides to birds and other terrestrial wildlife. The latter pesticides are generally highly toxic to birds in acute laboratory tests. None of the completed studies has successfully refuted the presumptions of acute risk to birds specified by the acute avian LOC. Most of the field studies have relied upon carcass searches as the primary investigational technique. A preliminary retrospective evaluation of results from the existing field studies evaluated by OPP confirms the prediction of bird mortality in the field from laboratory acute-toxicity data for OP and carbamate insecticides (see Table 1-3). A more detailed retrospective analysis of these studies is being planned through USEPA's Office of Research and Development (ORD).

Only 2 of these pesticides have been scrutinized through the entire regulatory review process, which also accounts for the benefits of pesticide use: diazinon for use on golf courses and sod farms and all major uses of granular formulations of carbofuran. Most of the remaining OP and carbamate insecticides are currently awaiting risk-versus-benefit decisions in the reregistration eligibility decision (RED) process. In addition, new insecticides are being submitted to the USEPA for registration each year. Some are new OPs and carbamates, but they are proposed for use at drastically lower label use rates. Others combine a new mode of action with lower use rates (Leicht 1993).

Defining and controlling the problem

History demonstrates that terrestrial pesticide field studies are not new with respect to data requirements and published guidance. Further, the literature is replete (dating back to at least 1970) with references to approaches, techniques, and examples for field study conduct (USEPA 1982, 1984, 1992a; Schemnitz 1980; Fite et al. 1988; Dingledine and Jaber 1990; Greig–Smith 1990; Kendall and Akerman 1992). From 1975 to 1992, USEPA shifted from requiring simulated small-pen avian studies to requiring Level 1 field studies for avian assessments. With the latest policy decisions (Fisher 1992), avian pesticide field study requests will generally not be required. USEPA will rely on laboratory data, the accumulated data from previous avian pesticide field studies, and extant bird-kill monitoring data to make regulatory decisions on avian risk (USEPA 1994a, 1994b) ear-

Table 1-3   Twenty terrestrial/pesticide field studies confirming acute risk to wildlife

Insecticides	Use/rate (lb/acre)	Estimated avian LD50/ft²	Animal
Granular			
Carbamate #1	4.95 ai/A	57.2	8 Birds 6 Mammals 3 Others
Carbamate #1	9.9 ai/A	114	15 Birds 1 Mammal
Carbamate #1	1.5 ai/A	114	12 Birds 3 Mammals 1 Other
Carbamate #1	3.0 ai/A	42	4 Birds 1 Other
Carbamate #1	2.25 ai/A	38	5 Birds
Carbamate #1	3.0 ai/A	300	28 Birds 9 Mammals
Carbamate #1	3.0 ai/A	300	4 Birds 2 Mammals
OP #1	3.0 ai/A	134	7 Birds 16 Mammals 2 Other
OP #2	1.3 ai/A	40 to 260	8 Birds 5 Mammals
OP #3	2.6 ai/A	8 to 14	20 Birds 14 Mammals 3 Others
Carbamate #2	3.7 ai/A	204	912 Birds
Carbamate #2	.075 ai/A	66	34 Birds
Carbamate #2	.075 ai/A	66	107 Birds
Carbamate #2	.075 ai/A	30	65 Birds
Carbamate #2	.075 ai/A	30	None

| | Table 1-3   continued | | |
Insecticides	Use/rate (lb/acre)	Estimated avian LD50/ft^2	Animal
Non-Granular			
OP #4	1.5 ai/A	70	59 Birds 109 Mammal 5 Others
OP #4	1.5 ai/A	70	5 Birds 24 Mammals
OP #5	1.5 ai/A	3.8	1 Impaired Bird
OP #6	0.5 ai/A	2.8	8 Birds
OP #6	0.2 ai/A	1.14	13 Birds

lier in the process. When LOCs are exceeded or bird-kill monitoring data indicate a concern, further regulatory action may be considered. This would include risk mitigation, i.e., procedures to reduce or mitigate the identified risk. Risk mitigation would be followed by post-registration, site-specific, residue- or ef-

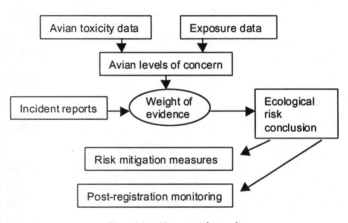

Figure 1-2   New eco-risk paradigm

fects- monitoring requirements to determine if the risk-mitigation procedures are effective (see Figure 1-2).

Avian pesticide field studies often have been viewed by OPP risk managers as problematic. They do not believe that these studies contribute greatly to the regulatory decision-making process (Fisher 1992). Reasons vary, but they include excessive cost; the time required for conduct, submission, and review of the avian field studies; lack of standardization on design and conduct; and inadequate guidance. As risk assessors, we need to better identify and agree upon the most critical risk-measurement and risk-assessment endpoints for these types of field studies. Assessment endpoints are "valued attributes of the environment" (Suter 1989) or "environmental values to be protected" (USEPA 1992b), e.g., a stable bird population. Measurement endpoints are measurable responses (e.g., bird mortality) to a stressor (e.g., a pesticide application) that are related to the assessment endpoint (Suter 1990). The USEPA has provided a framework for considering these endpoints in ERA (USEPA 1992b). In addition, it has identified ecological concerns from regulatory actions in the past and has made recommendations for improving ecological considerations in the future (USEPA 1994c). Only when these endpoints are clarified will scientists be able to design and OPP risk managers be able to use these field studies in FIFRA decisions regarding no unreasonable adverse effects.

What is the question?

Based upon recent policy changes, USEPA generally will not require avian pesticide field studies for making regulatory decisions. While field studies are still an option, USEPA does not plan to require these studies on a regular basis prior to making a regulatory decision on registration or reregistration. Situations where avian pesticide field studies may still be required include the following: when the pesticide under review exceeds an LOC and has a new mode of action; if it requires post-registration avian effects monitoring to evaluate risk mitigation measures; and if there are bird-kill incident reports (unspecified number) associated with its use. Previously, the avian pesticide field studies were designed to provide results that would answer the question "Can we determine that there will be no adverse effects on birds?" The answer was usually "No." That same question can be answered without avian pesticide field studies, but with less certainty. More frequently today, however, risk managers are asking other questions, some of which can best be answered with differently designed field studies. Salient questions might be "What are the magnitude, duration, and extent of the adverse effect?" and "How can we mitigate these effects short of canceling the registration of the pesticide use (implying a role for the risk assessor in developing practical risk-mitigation measures)?" Therefore, the policy change could, in some cases, result in avian pesticide field study requests to further characterize the risk of a new group of pesticides. This could occur either pre- or post-registration of the pesticide.

New responsibilities and opportunities

Risk managers prefer unequivocal data to make regulatory decisions. USEPA believes field studies do not provide such data. The field studies have offered evidence of bird mortality, but have not been satisfactory in quantifying the extent of effects. The risk assessors—the wildlife biologists and ecologists in regulatory agencies—however, bear the increasing burden of interpreting and communicating the results of laboratory and existing field studies in terms of all available data. They must characterize the risk in terms of the weight of the available evidence. This includes presenting the strengths, weaknesses, and uncertainties of the data. Increasingly, they are also called on to refine

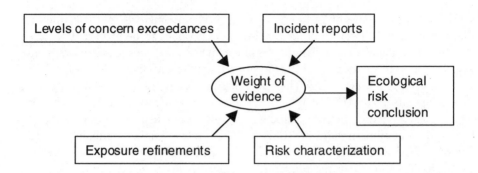

Figure 1-3   New eco-risk paradigm continued

exposure estimates and characterize risk by considering the following: species tested versus species exposed; species behavior as an indicator of pesticide exposure and effects; extent of exposure and use; quality, use, and value of treated habitat; and extrapolability of data. In addition, they are being asked to propose or at least comment on the likelihood that potential risk-mitigation measures will reduce the risk below the LOC (see Figure 1-3).

The risk assessor may also be required to answer questions about risk to avian populations—not comparing existing risk from hunting, natural causes, and other human activities, but describing it in terms of cumulative risk from the uses of the pesticide. Thus, there is a growing need for information on existing risk to bird populations. Further, the risk assessor must be prepared to make risk comparisons and be able to rank pesticides based on their potential risk. Ranking will provide perspective to the risk managers so they can set resource priorities and make environmentally protective regulatory decisions (USEPA 1992a).

Researchers also have a challenge. The current focus is on direct acute and chronic effects of pesticides on birds. Biologists recognize that the widespread use of pesticides that are acutely and chronically toxic to birds or their foodbase can result in an ecological impact beyond direct bird mortality. Other direct and indirect effects that are difficult to measure in the field can be even more important for sustaining local bird populations and their communities. For example, chick survival in the field may provide more realistic predictions of risk to upland game bird populations than predictions of bird mortality based upon laboratory values (Bennett 1993). Improved field-study designs and techniques, as well as advances in quantitative models, would provide helpful tools for risk assessors in characterizing avian risks at the population and community levels. While more quantitative avian population models that could be used by regulatory scientists are likely to be developed in the future (Kendall and Lacher 1992), improved techniques for avian field studies, such as radiotelemetry, are regularly being developed today (e.g., Rappole and Tipton 1991).

Does USEPA's new policy decision afford more or less protection to birds? Considering that only 2 highly toxic OP and carbamate insecticides (only 1 with major uses) have been rigorously regulated (canceled) in almost 20 years prior to these changes, the answer is probably the same or more. The policy change provides a window of opportunity for widespread use of risk reduction and mitigation for a large number of highly toxic pesticides.

Avian radiotelemetry—how can it improve risk assessment and regulatory decisions?

The real problem with avian pesticide field studies lies with the interpretation and use of results for the regulatory decision-making process. To improve interpretation of study results, we need to identify and better understand what key questions best facilitate the regulatory process and then design field techniques to address these questions.

Considering current limitations, Edwards (1990) states that "[Radiotelemetry] is potentially the most powerful technique available for use in wildlife trials measuring [wildlife] exposure and survival and improving the chance of [carcass] location." Other authors have focused on the utility of the technique in collecting data on the ways in which animals come in contact with the pesticide (Dingledine and Jaber 1990; Edwards 1990; Hart 1990; Kendall and Akerman 1992).

Hart (1990) makes the case that animals move, and rather than viewing this particular behavior as a hindrance, field studies should be designed with the proper techniques to investigate their movement. Hart goes so far as to say that "an understanding of animal movements is essential to good regulatory decisions." Bird movement in and out of treated fields has certainly hindered the interpretation of the results of avian pesticide field studies. Refined radiotelemetry methods for birds and better application to avian pesticide field studies can improve interpretation of the results of these studies. Clearer results will lead to improved communication of risk conclusions to risk managers, who should then provide more "avian risk conscious" regulatory decisions.

References

Bennett RS. 1993. *Corvallis Research Update*. 3(1):2. Corvallis OR: USEPA Environmental Research Laboratory.

Dingledine JV, Jaber MJ. 1990. A review of current field methods to assess the effects of pesticides on wildlife. In: Somerville L, Walker CH, editors. Pesticide effects on terrestrial wildlife. London: Taylor & Francis. p 241–255.

Edwards PJ. 1990. Assessment of survival methods used in wildlife trials. In: Somerville L, Walker CH, editors. Pesticide effects on terrestrial wildlife. London: Taylor & Francis. p 129–142.

[FIFRA] Federal Insecticide, Fungicide, and Rodenticide Act. 1988. Public Law 92–516, Title 7, United States Code. Washington DC: USEPA, Office of Pesticide Programs, PMSD, Information Services Branch.

Fisher LJ. 1992. Memorandum: decisions on the ecological, fate, and effects task force. Washington DC: USEPA, Office of Pesticides and Toxic Substances.

Fite EC, Turner LW, Cook NJ, Stunkard C. 1988. Guidance document for conducting terrestrial field studies. Washington DC: USEPA, Office of Pesticide Programs. EPA 540/09–88–109.

Greig–Smith PW. 1990. Intensive study versus extensive monitoring in pesticide field trials. In: Somerville L, Walker CH, editors. Pesticide effects on terrestrial wildlife. London: Taylor & Francis. p 217–239.

Hart ADM. 1990. Behavioral effects in field tests of pesticides. In: Somerville L, Walker CH, editors. Pesticide effects on terrestrial wildlife. London: Taylor & Francis. p 165–180.

Kendall RJ, Akerman J. 1992. Terrestrial wildlife exposed to agrochemicals: an ecological risk assessment perspective. *Environ Toxicol Chem* 11:1727–1749.

Kendall RJ, Lacher Jr TE, editors. 1992. Wildlife toxicology and population modeling: integrated studies of agroecosystems. Boca Raton FL: Lewis. 576 p.

Leicht W. 1993. Imidacloprid—a chloronicotinyl insecticide. *Pestic Outlook* 4(3):14–21.

Rappole JH, Tipton AR. 1991. New harness design for attachment of radio transmitters to small passerines. *J Field Ornithol* 62(3):335–337.

Schemnitz SD, editor. 1980. Wildlife management techniques manual, 4th Edition: Revised. Washington DC: The Wildlife Society. p 686.

Suter II GW. 1989. Ecological endpoints. In: Warren–Hicks W, Parkhurst BR, Baker Jr SS, editors. Ecological assessments of hazardous waste sites: a field and laboratory reference document. Washington DC: USEPA. EPA 600/3–890013.

Suter II GW. 1990. Endpoints for regional ecological risk assessments. *Environ Manage* 14(1):19–23.

[USEPA] U.S. Environmental Protection Agency. 1975. Guidelines for registering pesticides in the United States. 40 CFR (123), Part 162.

[USEPA] U.S. Environmental Protection Agency. 1982. Pesticide assessment guidelines, subdivision E, hazard evaluation: wildlife and aquatic organisms. Washington DC: USEPA. EPA–540/09–82–024.

[USEPA] U.S. Environmental Protection Agency. 1984. Data requirements for pesticide registration: final rule. 40 CFR (207), Part 158.

[USEPA] U.S. Environmental Protection Agency. 1992a. Comparative analysis of acute avian risk from granular pesticides. Office of Pesticide Programs. Washington DC: USEPA.

[USEPA] U.S. Environmental Protection Agency. 1992b. Framework for ecological risk assessment. Risk Assessment Forum. Washington DC: USEPA. EPA/630/R–92/001.

[USEPA] U.S. Environmental Protection Agency. 1994a. Ecological incident information system: user manual for the ecological incident information system. Office of Pesticide Programs. Washington DC: USEPA. EPA 734–K–94–001.

[USEPA] U.S. Environmental Protection Agency. 1994b. State contact personnel responsible for documenting and storing data on incidents involving pesticide exposures to wildlife. Washington DC: USEPA, Office of Pesticide Programs.

[USEPA] U.S. Environmental Protection Agency. 1994c. Managing ecological risks at EPA: issues and recommendations for progress. Center for Environmental Research Information, Office of Research and Development. Washington DC: USEPA. EPA/600/R–94/183.

Chapter 2

Ecotoxicological principles for avian field studies using radiotelemetry or remote sensing

Anne Fairbrother

Disclaimer: The information in this document has been funded by the U.S. Environmental Protection Agency (USEPA). It has been subjected to USEPA review and approved for publication.

This section provides a wildlife toxicologist's perspective of avian field studies and focuses on the types of information needed when conducting field studies to determine how a specific use of a single pesticide affects indigenous birds. Radiotelemetry is a tool by which one subset of wildlife ecotoxicological studies can be done, although it is also a powerful tool for determining how free-ranging animals come in contact with a variety of contaminated environments and what the short- and long-term effects might be. No attempt is made to limit the discussion to applications of extant technologies. Rather, the discussion centers around the information that we wish to gain and the context in which it will be used. By stating the needs of the end-user, the challenge to telemetry experts is to advance the engineering aspects of radiotelemetry and other remote sensing technologies. This should provide better tools for collecting ecologically relevant data in a cost-effective, real-time fashion.

Approach

The objective of a pesticide field study usually is to test the null hypothesis that application of a pesticide will have no effect on indigenous birds. Of course, this simple statement leads to many further questions: what is the ecological significance of "effect," what do we mean by "no effect," what do we know about dose–response relationships, does "application" refer to a single application or multiple applications, and what is the time period of concern? The focus here is not to address broad ecological concerns at this time, but to limit discussion to direct biological and toxicological responses of birds to a pesticide applied under a selected field regime. Within this context, the wildlife toxicologist is interested in gathering the following information: 1) what can we observe about birds' movements and behavior patterns that is likely to result in their exposure to the chemical; 2) what is the actual exposure of each individual and the population to the chemical; and 3) what are the physiological and behavioral consequences of chemical exposure (including mortality and reproductive endpoints) to the individual and population? Once we have this information, we can develop simple cause-and-effect relationships between pesticide application and bird responses. Given an appropriate study design, we can make 2 inferences: 1) how the observed responses vary depending upon the application rate (dose-response relationships) or the relative hazards of various compounds (comparative risk), and 2) differences in the magnitude of the effects on the various species present (comparative toxicology). The following are examples of the type of data we collect in order to develop these cause-and-effect relationships.

Location and movement of the population

Having 3-dimensional information about the location of the population of birds of interest within the study area in 3-dimensional space is vital. The 3 dimensions are latitude, longitude, and elevation. Latitude and longitude coordinates are needed in order to know when the birds are in the field where the pesticide is being applied. Multiple readings over time are needed to know when the birds are present and, therefore, have the potential to be exposed to the chemical. Spatial movement patterns can also provide information about what the bird is doing (flying, resting, feeding, nesting) which, in turn, provides keys to exposure potential as well as information about the effect of the pesticide application. Data about movement in the third dimension (elevation) can be used to augment information about certain behaviors (i.e., flying versus moving on the ground). Knowing the location of birds in real time is also useful for retrieving carcasses (from pesticide-induced mortality or other causes), for quantifying mortality, for diagnostic procedures, or for capturing live birds for sampling purposes. Special modifications (e.g., mortality switches, activity recorders, physiological-measurement devices) are useful but technologically difficult additions.

Behavior of the population

Once we know where the birds are located, we then need to know what they are doing there and if their behaviors are altered due to the pesticide application. In particular, we are interested in determining the effects of pesticide applications on those behaviors that will impact population demographics, i.e., mortality, natality, and dispersal. It is important to gather information about mortality in a time-dependent fashion. In other words, we wish to know when birds die as well as how many die. This will contribute to our ability to link mortality to the pesticide application, both by a time association and through various biochemical procedures (e.g., cholinesterase activity) that degrade with time after death. Ultimately, we calculate the survival rate rather than mortality rate, but these data are obviously related.

Reproduction of the population

Reproduction, the sum of a series of integrated behaviors, is another behavior to measure, and it should encompass the entire process: from mating behaviors through actual fertilization, egg incubation, and raising of young until they leave the nest. In strict ecological terms, we should include the period from fledging (leaving the nest) until the individuals mature and engage in mating behaviors themselves. However, the focus here is on the period from initiation of mating behaviors until the young fledge.

Mating behaviors in birds can be very complex, and birds frequently rely on visual cues. Conspicuousness or condition of plumage is a factor in mate selection that could be altered by pesticide exposure (Hamilton and Zuk 1982). Birds must initiate appropriate behaviors in the correct sequence and phenology and then must be able to respond appropriately to behaviors of conspecifics. Mating behavior has both learned and innate components, so it is not surprising that neurotropic pesticides have been shown to alter these patterns (Forsyth 1980; Grue and Shipley 1981; Busby et al. 1990). Once mating has occurred, egg production can be reduced due to pesticide exposure (Stomborg 1981; Rattner et al. 1982), as can proper formation of the egg and eggshell (Bennett and Bennett 1990; Scott et al. 1975). Information about fertility and eggshell quality generally is gathered in controlled laboratory studies, but it would be extremely useful to be able to gather similar data in the field to expand our knowledge about interspecific differences and the interactions of other environmental stressors (parasites, disease, extreme weather) with pesticide exposure. Nesting behaviors (amount of time spent sitting on eggs, how frequently the eggs are turned, whether the nest is incubated until term or abandoned early) have only occasionally been studied in relation to pesticide effects, (White et al. 1983; Brewer et al. 1988; Busby et al. 1990; Bennett et al. 1991) but they provide additional clues about cause-and-effect relationships relative to the final reproductive output.

Physiological effects (biomarkers) of the population
In addition to gathering data on pesticide effects on survival and reproduction, ecotoxicological studies would benefit from data concerning sublethal physiological effects. This area of study has largely been ignored due to the difficulty of obtaining the required information. Examples of questions that could be addressed within an ecological context include the following: l) does exposure to pesticide applications change the metabolic rates of the birds, thereby altering the amount of food and subsequent foraging time needed for survival or reproductive efforts (Dominguez et al. 1993); 2) does exposure to pesticides make birds more susceptible to concomitant environmental stressors such as disease and parasitism (Porter et al. 1984); and 3) do pesticides change hormone production patterns and interfere with growth and reproduction patterns or other more subtle behaviors (Bennett et al. 1991; Brewer et al. 1988)? Answers to these types of questions would help to establish definitive cause-and-effect relationships between pesticide application and subsequent changes in the local populations. More importantly, they would provide clues to the mechanisms by which the pesticides exert their effects on the birds. This would provide us with stronger predictive models about the ecotoxicological effects of pesticides and allow modifications of application regimes or new pesticide formulations to be made with a strong scientific basis. Remote sensing applications for monitoring physiological measurements will be discussed in more detail in another chapter.

Discussion

It is very important in field studies to know how much pesticide exposure each individual has received. Without this information, cause-and-effect relationships must be based on strong correlative relationships. There also are available surrogate measures (biomarkers) of exposure such as cholinesterase activity in brain and plasma (Mineau 1991) and chemical residue on crop and stomach contents. For some chemicals, we have direct exposure measurements through analysis of residues of parent compounds or breakdown products (e.g., oxones) in various tissues and excreta. Tissue residue studies are not ideal, however, as the current generation of pesticides frequently do not leave such residues. Those that do leave residues generally require destructive sampling of the animal and provide information only about the amount of chemical present at the time of sampling. Excreta measures of chemical content would be nondestructive and could be sampled repeatedly through time. Excreta collected from nests can be related to a specific individual (or two), but this restricts our ability to measure what is occurring in non-nesting birds or during the portion of the day when the bird is not in the nest. Merely locating the individual coincident with pesticide application in time and space does not provide a measure of exposure. For example, foliage can provide shelter from exposure while preening activities can increase total exposure. From an ecotoxicological point of view, exposure assessment is the weakest part of pesticide risk-assessment studies.

Summary

Ecotoxicological studies for pesticide risk assessments strive to develop cause-and-effect relationships between pesticide application and adverse effects on birds (both as individuals and as populations) and to determine the mechanisms by which the observed effects occur. In order to accomplish this, data are collected to determine whether the location of the birds puts them into potential contact with the pesticide, how the pesticide changes behaviors and the physiology of the animals, and how these changes result in increased mortality and reduced reproduction. Radiotelemetry and other remote sensing technologies can be powerful tools to aid in gathering data. We need to extend the current capabilities of these tools, while providing information in real time and in a user-friendly fashion.

References

Bennett JK, Bennett RS. 1990. Effects of dietary methyl parathion on northern bobwhite egg production and eggshell quality. *Environ Toxicol Chem* 9:1481–1485.

Bennett RS, Williams BA, Schmedding DW, Bennett JK. 1991. Effects of dietary exposure to methyl parathion on egg laying and incubation in mallards. *Environ Toxicol Chem* 10:501–507.

Brewer LW, Kendall RJ, Driver CJ, Zenier C, Lacher Jr TE. 1988. Effects of methyl parathion on ducks and duck broods. *Environ Toxicol Chem* 7:375–379.

Busby DG, White LM, Pearce PA. 1990. Effects of aerial spraying of fenitrothion on breeding white-throated sparrows. *J Appl Ecol* 27:743–755.

Dominguez SE, Menkel JL, Fairbrother A, Williams BA, Tanner RW. 1993. The effect of 2,4-dinitrophenol on metabolic rate of bobwhite quail. *Toxicol Appl Pharmacol* 123:226–233.

Forsyth DJ. 1980. Effects of dietary fenitrothion on the behavior and survival of captive white-throated sparrows. In: Varty IW, editor. Environmental surveillance in New Brunswick, 1978–1979. Effects of spray operations for forest protection against spruce budworm. Committee for Environmental Monitoring of Forest Insect Control Operations, Dept. For. Res., Univ. New Brunswick, Fredericton, Canada. p 27–28.

Grue CE, Shipley BK. 1981. Interpreting population estimates of birds following pesticide applications—behavior of male starlings exposed to an organophosphate pesticide. *Stud Avian Biol* 6:292–296.

Hamilton WD, Zuk M. 1982. Heritable true fitness and bright birds: a role for parasites? *Science* 218:384–387.

Mineau P. 1991. Cholinesterase-inhibiting Insecticides: Their Impact on Wildlife and the Environment. Amsterdam, The Netherlands: Elsevier. 348 p.

Porter WP, Hinsdill R, Fairbrother A, Olson LJ, Jaeger J, Yuill TM, Bisgaard S, Hunter WG, Nolan K. 1984. Toxicant-disease-environment interactions associated with suppression of immune system, growth, and reproduction. *Science* 224:1014–1017.

Rattner BA, Sileo L, Scanes CG. 1982. Oviposition and the plasma concentration of LH, progesterone and corticosterone in bobwhite quail (*Colinus virginianus*) fed parathion. *J Reprod Fertil* 66:147–155.

Scott M L, Zimmermann JR, Marinsky S, Mullenhoff PA, Rumsey GL, Rice RW. 1975. Effects of PCBs, DDT, and mercury compounds upon egg production, hatchability, and shell quality in chickens and Japanese quail. *Poult Sci* 54:350–368.

Stromborg KL. 1981. Reproductive tests of diazinon on bobwhite quail. In: Lamb DW, Kenaga EE, editors. Avian and mammalian wildlife toxicology: 2nd conference. ASTM STP757. Philadelphia PA: American Society for Testing and Materials. p 19–30.

White DH, Mitchell CH, Hill EG. 1983. Parathion alters incubation behavior of laughing gulls. *Bull Environ Contam Toxicol* 31:93–97.

Chapter 3

Radiotelemetry in wildlife biology: a brief overview

Larry W. Brewer

Cochran and Lord (1963), Marshall and Kupa (1963), Southern (1965), Marshall (1965), and Craighead and Craighead (1965) were among the early users of radiotelemetry to monitor the activities of free-ranging animals. The Craigheads pointed out that biotelemetry extended the range of the observer's powers to include continuous monitoring of radio-tagged animals both day and night, and they predicted that its use (radiotelemetry) would become a serious method of data collection in future studies. Perhaps 30 years ago no one, including these insightful authors, envisioned just how significant the radio tracking of wildlife would become in the various wildlife science disciplines.

The first documentation of radio-transmitter use on wildlife involved the monitoring of chipmunk heart rates using an implanted transmitter following methodology developed by the U. S. Navy to monitor test pilot heart rates (LeMunyan et al. 1959). Shortly thereafter, a publication describing the use of an externally mounted transmitter to monitor heart rates and wing beats of mallards was also published (Eliassen 1960). As pointed out by Kenward (1987), these technological advances were the result of the development of the transistor. Also, these early transmitters could only produce a continuous signal for which changes in signal frequency were indicators of physiological changes. Before wildlife radio tracking could become practical, the transmitter battery life had to be extended substantially. This was accomplished by converting the continuous signal to a pulse rather than a constant signal. This pulsing of the transmitter signal increased battery life by many-fold, depending on the time between pulses. This advance took place in the early 1960s, which was the time that field radio tracking of free-ranging animals was initiated.

During the past 3 decades radio tracking of wildlife has proven to be one of the most significant advances in the field of wildlife biology. In the earlier years, radio tracking was limited to larger animals due to the size of the transmitter packages. With the advent of micro-electronics and miniaturization of radio transmitters, we can now radio tag passerine species and mammals as small as a deer mouse. Radio-tracking research has been conducted with numerous animal species of all sizes from polar bears to American robins, and it has allowed biologists to measure animal home range size, migration routes, movement patterns, survival rates, and other behavioral variables. The published literature for wildlife radio tracking is voluminous. Kenward (1987) suggests that the following are very useful beginning literature sources: Amlaner and MacDonald (1980), Cheeseman and Mitson (1982), and all of the North American proceedings of the International Conferences on Wildlife Biotelemetry. Additionally, 2 books provide comprehensive handling of wildlife telemetry methodology and techniques for analyzing telemetry data for beginners and experienced individuals: 1) *Wildlife Radio Tagging* by Kenward (1987), and 2) *Analysis of Wildlife Radio-Tracking Data* by White and Garrott (1990).

Since the first radio-tracking studies, telemetry equipment manufacturers have continuously worked toward further miniaturization of transmitters and extending transmitter life with lighter weight batteries. This developmental effort is driven by the need to track smaller animals for longer periods of time. An additional scientific need is to monitor animals continuously 24 hours per day. To accomplish this, fully automated telemetry systems have been proposed and there have been several automated telemetry systems designed and used in the field with varying degrees of success (Cochran et al. 1965; Craighead and Craighead 1965; Marshal 1965; Kenward 1987; Larkin et al. 1996). LOTEK Engineering, Inc., Newmarket, Ontario, has recently designed and is marketing an automated system that has proven effective in tracking movements of anadromous fish in rivers. McQuillen et al. (1996) report on the development of an automated telemetry system designed specifically to monitor the presence and absence of birds on agricultural fields, which is relevant to pesticide registration studies. Several manufacturers now offer wireless automated radio-tracking systems that are based on satellite communications similar to geographic positioning systems. These are currently limited to use on larger animals due to the size of the electronic components of the transmitter.

Pesticide registration field studies have unique requirements. The U.S. Environmental Protection Agency has routinely required use of telemetry in 2 circumstances: 1) assessment of field risk to nontarget animals and 2) field efficacy for vertebrate pesticides. For field efficacy studies, transmitters must be small enough to use on starlings or ground squirrels, have 30 d of battery life and provide a mortality signal. For field risk assessment, a fully automated system that can provide accurate transmitter location and can be used on very small animals, such as sparrow-sized birds, is the ultimate tool needed to obtain temporal patterns of site use by individual animals. This would allow for accurate estimates of exposure for those animals following a pesticide application on a specific agricultural field as well as definitive survival analysis. While such a system is not fully operational at this time, some of the current work on automated systems referenced above (McQuillen et al. 1996) appears very close to achieving this ability. As the radiotelemetry and radio-tracking technology continues to improve, its utility in wildlife science disciplines will likely expand.

References

Amlaner CJ, Macdonald DW. 1980. A practical guide to radio tracking. In: Amlaner CJ, Macdonald DW, editors. A handbook on biotelemetry and radio tracking. Oxford UK: Pergamon. p 143–159.

Cheeseman CL, Mitson RB, editors. 1982. Telemetric studies of vertebrates. Symposium of the Zool. Soc. of London, 49. London: Academic.

Cochran WW, Lord RD. 1963. A radio-tracking system for wild animals. *J Wildl Manage* 27: 9–24.

Cochran WW, Warner DW, Tester JR, Kuechle VB. 1965. Automatic radio tracking system for monitoring animal movements. *Bioscience* 15: 98–100.

Craighead Jr FC, Craighead JJ. 1965. Tracking grizzly bears. *Bioscience* 15: 88–92.

Eliassen E. 1960. A method for measuring the heart rate and stroke/pulse pressures of birds in normal flight. Årbok Universitiet Bergen, *Matematisk Naturvitenskapelig* 12:1–22.

Kenward R. 1987. Wildlife radio tagging. New York: Academic. 222 p.

Larkin RP, Raim A, Diehl RH. 1996. Performance of a non-rotating direction-finder for automatic radio tracking. *J Field Ornithol* 67: 59–71.

LeMunyan CD, White W, Nybert E, Christian JJ. 1959. Design of a miniature radio transmitter for use in animal studies. *J Wildl Manage* 23: 107–110.

Marshall WH. 1965. Rufffed grouse behavior. *Bioscience* 15: 92–94.

Marshall WH, Kupa JJ. 1963. Development of radio-telemetry techniques for ruffed grouse studies. *Trans of North Am Wildl and Nat Resour Conf* 28: 443–456.

McQuillen HL, Sullivan JP, Brewer LW. 1996. Use of a prototype automated radio telemetry system to monitor bird presence and survival in or near agricultural fields. Proc. of Forum on Wildl. Telemetry Innovations, Evaluations and Research Needs, Snowmass, CO. 21–23 Sept 1997 (in prep.). Jamestown MD:USGS, Biological Resources Division.

Southern WE. 1965. Avian navigation. *Bioscience* 15: 87–88.

White GC, Garrott RA. 1990. Analysis of wildlife radio-tracking data. New York: Academic. 383 p.

Chapter 4

Radiotelemetry to determine exposure and effects of organophosphorus insecticides on sage grouse

Lawrence J. Blus and John W. Connelly

This chapter discusses the history of die-offs of sage grouse from organophosphorus (OP) pesticides in southeastern Idaho cropland. These die-offs and subsequent collections of birds around treated fields led to a study of effects of OP insecticides on sage grouse that were captured and fitted with transmitters in 1985 and 1986. Of 82 radio-marked grouse, at least 18% were in cropland at the time of spraying with the OP insecticides dimethoate and methamidophos, 17% became seriously intoxicated, and 11% died from poisoning by these 2 OPs, as determined by inhibition of brain cholinesterase (ChE) activity and residues of these compounds in ingesta. One large die-off occurred in 1986 when an alfalfa field was sprayed with dimethoate; about 200 sage grouse, including 7 radio-marked birds, occupied the field at the time of spraying; and 63 dead birds were found during 12 d after treatment. Residues in ingesta and assays of ChE activity in brains of 43 grouse indicated that dimethoate was responsible for the die-off. There was 65% mortality of 31 sage grouse initially radio marked when intoxicated from dimethoate. Proposed research will provide a more realistic indication of the risk of OPs to the sage grouse population as a whole by trapping birds on leks at various distances from cropland, by tracing survival over several years, and by using a proportional hazards model.

Replacement of long-lived organochlorine (OC) insecticides with the shorter-lived anticholinesterase (antiChE) OP and carbamate compounds generally reduced persistent problems on wildlife populations. Nevertheless, different modes of action of antiChE compounds have had negative effects, particularly from a short-term perspective where acute or subacute toxicity (Hill and Fleming 1982; Grue et al. 1983; Henny et al. 1985; White et al. 1990) and reductions in arthropod prey populations (Rands 1985; Potts 1986) are major concerns.

Die-offs of sage grouse *(Centrocercus urophasianus)* in agricultural fields in southeastern Idaho beginning in 1981 led to a radiotelemetry study in 1985 and 1986 where effects of OPs on sage grouse were documented (Blus et al. 1989). Our objectives here are to review the documented effects of OPs on sage grouse, offer new perspectives on these findings, and discuss the need for and probable success of continued ecotoxicological studies of radio-marked sage grouse.

Southeastern Idaho sage grouse study, 1985 to 1986

Study area and methods

The study areas, methods, and materials are described in detail elsewhere (Blus et al. 1989). Briefly, sage grouse, primarily females and young, were captured by night-lighting (Giesen et al. 1982) in sagebrush (*Artemisia* spp.) near cropland in July 1985 and 1986. Eighty-two grouse (39 in 1985 and 43 in 1986) were fitted with radio collars and followed through the summer to termination of the field season (23 August 1985 and 3 September 1986). We also attached a transmitter to grouse found intoxicated after exposure to OPs. Brains of grouse found dead were assayed for ChE activity (Ellman et al. 1961; Hill and Fleming 1982). We also used a standard for ChE each day that samples were assayed. Exposure to an antiChE compound was indicated when activity was < 2 SD of the control mean which was derived each year from hunter-killed and road-killed birds from non-agricultural areas. AntiChE exposure is postulated as the cause of death with inhibition 50% (Ludke et al. 1975; Hill and Fleming 1982). Crop or gizzard contents of sage grouse collected or found dead were analyzed with a gas chromatograph equipped with an electron capture detector; 10% of the residue analyses were confirmed with a mass spectrometer (Blus et al. 1989).

Survival functions of radio-marked sage grouse were estimated with the Kaplan-Meier (product limit) nonparametric estimator (Lee 1980) with a staggered entry scheme to preserve the relationship between the survival function and the calendar date (Pollock et al. 1989). Using this method, we estimated the probability of grouse surviving beyond a specified time to a specified date or number of days since marking. A chi-square test was used to compare survival of adult and juvenile grouse that were radio marked either when apparently healthy or when intoxicated.

Movements of sage grouse

The basic movement pattern of radio-marked sage grouse was to feed in cropland and to roost in sagebrush; 85% of marked grouse were recorded in cropland. Grouse tended to occupy crop fields nearest the area where they were trapped. Maximum movements of sage grouse from sagebrush into cropland were 2.3 and 3.9 km in 1985 and 1986, respectively; these grouse utilized cropland for several weeks. By late August, grouse tended to move farther into sagebrush and to make less use of cropland.

Generally, sage grouse in southeastern Idaho are migratory (Dalke et al. 1963; Connelly et al. 1988; Wakkinen 1990); movement to summer range, including cropland, begins in June. Maximum movement of adult sage grouse from winter to summer range was 82 km (Connelly and Markham 1983; Connelly et al. 1988). Distances moved from the nest to summer range by females with broods ranged from 3 to 21 km; 82% of sage grouse trapped and marked on leks subsequently moved to irrigated cropland (Gates 1983). Based on results of this study and others, most of the sage grouse breeding in relatively xeric (20 to 26 cm of precipitation) sagebrush habitats in southeastern Idaho use cropland for summer range; use of cropland increases sharply during extended periods of hot and dry weather. In southeastern Idaho, spraying of crops with pesticides is initiated in late spring, but most applications occur in July and August at the peak of cropland use by sage grouse. Under these conditions, there is potential for a large segment of the sage grouse population in southeastern Idaho to be exposed to pesticides; such exposure may contribute to population declines of this species and other gallinaceous game birds.

Exposure, intoxication, and mortality

Because grouse occupied alfalfa and potato fields during the time of heaviest applications of pesticides, many birds were exposed to the sprays. This exposure resulted in intoxication and death of a number of birds. In 1985, 6 of 39 radio-marked grouse (15%) became intoxicated after the alfalfa field they occupied was sprayed with the OP dimethoate; 2 of these died with 62 and 73% inhibition of brain ChE activity compared with the mean value obtained from control samples. The intoxicated birds could not walk or fly; they were emaciated, had diarrhea, frequently salivated, and sometimes uttered faint vocalizations. These signs are characteristic of poisoning by antiChE compounds such as OPs and carbamates. The biochemical lesion is phosphorylation or carbamylation of acetylcholinesterase and resultant accumulation of acetylcholine that disrupts nerve transmissions. The sequence of poisoning is inhibition of ChE; acetylcholine accumulation; disruption of nerve function, either centrally or peripherally; respiratory failure; and death by asphyxia (O'Brien 1960). Sublethal inhibition of ChE is reversible; 4 of the intoxicated grouse recovered after approximately 1 week. These birds appeared normal when they were shot 9 or 18 d postspray, but their brain ChE activity was still inhibited 31 to 35%.

While searching for intoxicated radio-marked grouse in alfalfa in 1985, we encountered 3 unmarked, intoxicated grouse. Brain ChE activity was inhibited 66 and 67% in 2 of these grouse that were euthanatized. The third intoxicated grouse recovered after it was radio marked, but it had 37% inhibition of brain ChE activity when shot 18 d postspray (Table 4-1).

In 1986, we observed approximately 100 sick or dead grouse around 3 alfalfa fields that were sprayed with dimethoate. The major die-off occurred on 1 August when an alfalfa field that contained a flock of about 200 sage grouse (including 7 radio-marked birds) was sprayed with dimethoate. About 30 intoxicated or dead grouse were observed on 2 August, and the last verified mortality was recorded on 12 August. We saw at least 2 grouse fall to the ground from flight. We found 63 dead grouse in the alfalfa field, including 5 of the 7 radio marked when healthy, 20 of 31 radio marked when intoxicated, and 38 birds without radios. Of the 31 grouse radio marked when intoxicated, 20 (65%) apparently died from dimethoate; 10 of these deaths were verified by brain ChE assays (Table 4-1). Most of the sick grouse attempted to move from alfalfa into sagebrush that bordered the field. Most grouse died in or at the edge of the alfalfa field, but 2 grouse radio marked when intoxicated died in sagebrush 0.8 and 1 km from the field border.

Table 4-1   Brain chloinesterase activity of sage grouse controls compared to birds collected or found dead in summer in or near southeastern Idaho cropland, 1985 and 1986 (from Blus et al. 1989)

Year	n	OP[b]	Condition	% Change from Control[a] Mean	Range	% of Grouse Exposed[c]	With ≥ 50% inhibition
1985	2	DI	Dead	−67	−73 to −62	100	100
	3	−	Dead[d]	+1	−10 to +14	0	0
	5	DI	Shot	−34	−37 to −31	100	0
	11	−	Shot	+7	−61 to +38	9	9
	2	DI	Sick	−67	−67 to −66	100	100
1986	43	DI	Dead	−74	−90 to −51	100	100
	2	ME	Dead	−41	−43 to −39	100	0
	8	DI	Shot	−14	−30 to + 6	25	0
	1	−	Dead[e]	−8		0	0

[a] Results of control ChE assays (mean micromoles of substrate [acetylcholine iodide] hydrolyzed/min/g of brain tissue ± 2 SD) were 12.5 ± 2.2 for 11 birds in 1985 and 15.3 ± 3.3 for 7 birds in 1986
[b] Exposure to DI dimethoate or ME methamidophos; − no known exposure
[c] Less than control mean 2 SD
[d] Includes roadkill, predator kill, and undetermined cause of death
[e] Roadkill

Avian and mammalian predators were attracted to the dead and dying grouse; 17 depredated grouse carcasses were found within 2 weeks postspray.

Assays of brains of 43 of the 63 sage grouse found dead (9 of the 43 were depredated) in the alfalfa field sprayed 1 August revealed 51 to 90% inhibition of ChE activity (Table 4-1). Brain ChE activity was depressed from 51 to 86% in the 9 depredated grouse. Nine grouse were radio marked when intoxicated and subsequently recovered: 5 shot on 3 September had brain ChE activity inhibited from 9 to 30%, 3 shot on 17 September had brain ChE activity that ranged from 87 to 106% of control values, and we lost the signal of the ninth bird (Table 4-1). The major differences in the die-offs from dimethoate in 1985 and 1986 were the number of birds involved and the time to the initial deaths. Also, none of the birds that died from dimethoate in 1985 had food in their crops; the gizzard contents of one contained a trace of dimethoate (0.02 μg/g). Crops of most grouse found dead the day after spray in 1986 contained ingested food, primarily alfalfa; residues of dimethoate ranged from 3 to 30 μg/g.

In August 1986, 2 of 12 sage grouse radio marked when apparently healthy near potato fields were found buried in or near a potato field the day after it was sprayed with methamidophos. The carcasses were partially eaten and then buried by coyotes *(Canis latrans)*. Brain ChE activity of the 2 grouse was depressed 39 and 43%, and crop contents of 1 grouse contained 18 µg/g methamidophos; these were the only sage grouse suspected of dying of OPs that had < 50% inhibition of brain ChE activity. Experimental evidence suggests that deaths of these two grouse resulted from exposure to methamidophos. For example, Japanese quail *(Coturnix japonica)* were critically intoxicated when euthanatized l h after receiving an oral dose of the OP dicrotophos; however, brain ChE activity was inhibited only about 40% (Hill 1989). Several experimental mallards killed by methamidophos had about 40% inhibition of brain ChE activity (CE Grue, personal communication).

The die-offs from dimethoate described here were apparently the first verified records of wildlife losses from this OP; we found only one other record of a die-off of birds from an application of methamidophos (Smith 1987). Factors that increased risk of dimethoate to sage grouse were their use of alfalfa fields for roosting, loafing, and feeding on sprayed alfalfa foliage. The conditions associated with methamidophos applications to potatoes that resulted in risk to sage grouse were similar to those associated with dimethoate applications to alfalfa. The crops of grouse shot or found dead in potato fields contained foliage of weeds and small amounts of insect material; however, sage grouse may occasionally eat potato foliage.

In other studies, the half-life of dimethoate and methamidophos on plants was < 4 d; however, residues of these systemic insecticides may persist for several weeks (Szeto et al. 1984; Westcott et al. 1987). Thus, intoxicated sage grouse in cropland may be reexposed to the OPs when ChE reversal is initiated and the grouse resume feeding on contaminated foliage or move to other recently treated fields.

Sublethal depression of ChE activity in the brain was not demonstrated to have persistent physiological effects in experimental birds in earlier studies (Metz 1958; Glow and Rose 1966; Banks and Russell 1967), but more recent studies presented evidence that certain OPs are capable of inducing long-term effects (Farage-Elawar and Francis 1987, 1988). In chickens, methamidophos had irreversible effects on neurotoxic target esterase that resulted in OP-induced delayed polyneuropathy (Johnson et al. 1989). These effects have not been noted in wild birds, but without a properly designed radiotelemetry study, there is essentially no possibility of identifying long-term effects.

The approximate time required for reactivation of ChE activity in brains of sage grouse intoxicated from dimethoate in this study was similar to the 26-d recovery period (from 55 to 64% inhibition to within 2 SD of the control mean) in experimental birds (Fleming and Grue 1981); however, additional research is required to more closely define this period.

Survival analysis

Survival analysis of the 39 sage grouse radio marked when apparently healthy in 1985 indicated that the probability of these grouse dying during the 45-d tracking period was 0.25 (mortality = 1 - survival); however, only 2 of 9 documented deaths were related to dimethoate intoxication (probability of dying from OPs = 0.10). Signals from 17 of the 39 grouse were lost before the study ended because of the short range of the transmitters (< 1.3 km) and related problems; thus, the mortality values were minimal estimates with low precision.

Ten of 43 sage grouse that were fitted with transmitters while healthy in 1986 died within 72 d with an overall mortality rate of 0.32. Of grouse radio marked when healthy, 7 juveniles died from OPs (5 from dimethoate and 2 from methamidophos), and 3 other birds (2 juveniles and an adult female) were depredated from 15 August to 17 September. The probability of a grouse dying from OP poisoning was 0.25. Although the last 3 birds that died spent 3 to 20 d after spraying in cropland, there was no evidence of their exposure to OP sprays, and their brains were not available for ChE assays.

The probability of mortality for 31 grouse, radio marked when found intoxicated in 2 alfalfa fields from 25 July (2 birds) up to 7 August 1986 (29 birds), was 0.76 up to 12 August when the last OP mortality from the second incident was verified and 0.78 to 3 September when several were collected by shooting. Dimethoate apparently accounted for deaths of 20 of the 31 grouse; ChE activity was inhibited > 50% in brains of each of 10 birds suitable for assay, and residues of dimethoate were detected in ingesta of some birds. All 8 intoxicated grouse that were radio marked the day following the second spray incident on 1 August died within 4 d; 12 of 21 grouse that were fitted with transmitters when intoxicated on 3 to 7 August died between 4 and 12 August. The grouse that died on 12 August was depredated; however, brain ChE activity was inhibited 55%. The longer range (2.0 to 2.5 km) of the transmitters used in 1986 improved efficiency of tracking compared to 1985; nevertheless, signals of 5 grouse were lost before the end of the study.

Age effects

Concerning sage grouse radio marked when apparently healthy in 1986, 7 of 28 juveniles died compared to 0 of 15 adults (P < 0.05) (Table 4-2). There was no significant difference (P > 0.05) in survival of adults and juveniles fitted with transmitters when intoxicated; however, all 5 adults died compared with 14 of 25 juveniles. Two adults were among 38 unmarked birds that probably died from dimethoate; however, sex and age were not determined for 13 birds.

Conclusions

Findings in 1985–1986 suggest that OPs may adversely affect sage grouse populations; however, we only considered those birds whose summer range included cropland. The mortality rate and sublethal intoxication of our marked population, induced by OPs and possibly other pesticides, was probably underestimated because: 1) sage grouse were

Table 4-2   Incidence of organophosphorus pesticide-related mortality of sage grouse by age and sex, southeastern Idaho, 1986 (from Blus et al. 1989)

Condition	Female		Male		Unknown sex	Unknown sex and age
	Adult	Juvenile	Adult	Juvenile	Juvenile	
Radiomarked						
Healthy	11 (0)[a]	9 (3)	4 (0)	9 (4)	10 (0)	
Intoxicated	4 (4)	11 (5)	1 (1)	14 (9)		1 (1)
Not marked						
Dead	(1)	(7)	(1)	(13)	(3)	(13)

[a] Number of grouse radiomarked with the number dying from OPs in parentheses

radio tracked during only part of the season when OPs and other pesticides were applied, 2) signals were lost from a number of grouse before the study terminated each year, 3) transmitters were removed from 5 healthy birds in 1986 and placed on intoxicated grouse, 4) some unrecorded exposure of marked grouse may have occurred between radio locations because the birds were not tracked continuously, and 5) delayed effects may have occurred. Circumstantial evidence indicates that the problem may be continuing in southeastern Idaho; an unverified report in 1989 indicated that a large number of sage grouse died at a cropland-sagebrush interface (Justin Naderman, personal communication). One objective of the workshop is to formulate or refine study designs for using radiotelemetry to investigate effects of pesticides on birds; therefore, the remainder of this book is directed toward that objective.

Although declines in the sage grouse population in southeastern Idaho are likely caused by habitat deterioration and adverse weather, population declines may be exacerbated by use of insecticides. Comparative toxicological data of OP and carbamate pesticides for gallinaceous birds indicate a wide spectrum of LD50 estimates for this order (Hudson et al. 1984). Evidence suggests that sage grouse are unusually sensitive to antiChE compounds; additional work is required to clarify this relationship.

This study, to be conducted over 3 years and involving the radio marking and monitoring of a large subsample of the sage grouse population, has the potential for generating data necessary to estimate the potential effects that pesticides have on sage grouse summering near croplands. A likelihood model for avian exposure data will be developed during the course of the study to depict the relationship between exposure to the pesticides and population recruitment and survival. Ultimately, the information obtained from this study may be useful for regulatory purposes to evaluate pesticide management practices in agricultural areas in southeastern Idaho and possibly in other areas.

Acknowledgments - We thank the many individuals who assisted with work in the laboratory and field, including J. Skalski for developing the model and C. Bunck, D. Sparling, and D. White for reviewing the manuscript.

References

Banks A, Russell RW. 1967. Effects of chronic reductions in acetylcholinesterase activity on serial problem-solving behavior. *J Comp Physiol Psychol* 64:262–267.

Blus LJ, Staley CS, Henny CJ, Pendleton GW, Craig TH, Craig EH, Halford DK. 1989. Impact of organophosphorus insecticides on sage grouse in southeastern Idaho. *J Wildl Manage* 53:1139–1146.

Connelly JW, Browers HW, Gates RJ. 1988. Seasonal movements of sage grouse in southeastern Idaho. *J Wildl Manage* 52:116–122.

Connelly JW, Markham OD. 1983. Movements and radionuclide concentrations of sage grouse in southeastern Idaho. *J Wildl Manage* 47:169–177.

Dalke PD, Pyrah DB, Stanton DC, Crawford JE, Schlatterer EF. 1963. Ecology, productivity, and management of sage grouse in Idaho. *J Wildl Manage* 27:811–841.

Ellman GL, Courtney KD, Andres Jr V, Featherstone RM. 1961. A new and rapid colorimetric determination of acetylcholinesterase activity. *Biochem Pharmacol* 7:88–95.

Farage–Elawar M, Francis RM. 1987. Acute and delayed effects of fenthion in young chicks. *J Toxicol Environ Health* 21:455–469.

Farage–Elawar M, Francis RM. 1988. Effects of multiple dosing of fenthion, fenitrothion, and desbromoleptophos in young chicks. *J Toxicol Environ Health* 23:217–228.

Fleming WJ, Grue CE. 1981. Recovery of cholinesterase activity in five avian species exposed to dicrotophos, an organophosphorous pesticide. *Pestic Biochem Physiol* 16:129–135.

Gates RJ. 1983. Sage grouse, lagomorph, and pronghorn use of a sagebrush grassland site on the Idaho National Engineering Laboratory. M.S. Thesis, Montana State Univ., Bozeman. 135 p.

Giesen KM, Schoenberg TJ, Braun CE. 1982. Methods for trapping sage grouse in Colorado. *Wildl Soc Bull* 10:224–231.

Glow PH, Rose S. 1966. Cholinesterase levels and operant extinction. *J Comp Physiol Psychol* 61:165–172.

Grue CE, Fleming WJ, Busby DG, Hill EF. 1983. Assessing hazards of organophosphate pesticides to wildlife. *Trans North Am Wildl Nat Resour Conf* 48:200–220.

Henny CJ, Blus LJ, Kolbe EJ, Fitzner RE. 1985. Organophosphate insecticide (famphur) topically applied to cattle kills magpies and hawks. *J Wildl Manage* 49:648–658.

Hill EF. 1989. Divergent effects of postmortem ambient temperature on organophosphorus- and carbamate-inhibited brain cholinesterase activity in birds. *Pestic Biochem Physiol* 33:264–275.

Hill EF, Fleming WJ. 1982. Anticholinesterase poisoning in birds: field monitoring and diagnosis of acute poisoning. *Environ Toxicol Chem* 1:27–38.

Hudson RH, Tucker RK, Haegele MA. 1984. Handbook of toxicity of pesticides to wildlife. U.S. Fish Wildl. Serv. Resour. Publ. 153. 90 p.

Johnson MK, Vilanova E, Read DJ. 1989. Biochemical and clinical tests of the delayed neuropathic potential of some O–alkyl–O–dichlorophenyl phosphoramidate analogues of methamidophos (O,S–dimethyl phosphorothioamidate). *Toxicology* 54:89–100.

Lee ET. 1980. Statistical methods for survival data analysis. Belmont CA: Lifetime Learning Publ. 557 p.

Ludke JL, Hill EF, Dieter MP. 1975. Cholinesterase (ChE) responses and related mortality among birds fed ChE inhibitors. *Arch Environ Contam Toxicol* 3:1–21.

Metz B. 1958. Brain acetylcholinesterase and a respiratory reflex. *Am J Physiol* 192:101–105.

O'Brien RD. 1960. Toxic phosphorus esters. New York: Academic. 434 p.

Pollock KH, Winterstein SR, Bunck CM, Curtis PD. 1989. Survival analysis in telemetry studies: the staggered entry design. *J Wildl Manage* 53:7–15.

Potts GR. 1986. The partridge. London, UK: Collins. 274 p.

Rands MRW. 1985. Pesticide use on cereals and the survival of grey partridge chicks: a field experiment. *J Appl Ecol* 22:49–54.

Smith GJ. 1987. Pesticides used and toxicology in relation to wildlife: organophosphorus and carbamate compounds. U.S. Fish Wildl. Serv. Resour. Publ. 170. 171 p.

Szeto SY, Mackenzie JR, Brown MJ. 1984. Disappearance of dimethoate, methamidophos and pirimicarb in lettuce. *J Environ Sci Health* B19:225–235.

Wakkinen WL. 1990. Nest site characteristics and spring–summer movements of migratory sage grouse in southeastern Idaho. M.S. Thesis, Univ. of Idaho, Moscow. 57 p.

Wakkinen WL, Reese KP, Connelly JW. 1992a. Sage grouse nest locations in relation to leks. *J Wildl Manage* 56:381–383.

Wakkinen WL, Reese KP, Connelly JW, Fischer RA. 1992b. An improved spotlighting technique for capturing sage grouse. *Wildl Soc Bull* 20:425–426.

Westcott ND, Lee YW, McKinlay KS. 1987. Persistence of dimethoate on wheat herbage and sweetclover herbage. *J Environ Sci Health* 22B:379–390.

White DH, Seginak JT, Simpson RC. 1990. Survival of northern bobwhites in Georgia: cropland use and pesticides. *Bull Environ Contam Toxicol* 44:73–80.

Chapter 5

Avian radiotelemetry techniques for pesticide registration field studies

Larry W. Brewer, Jeremy A. Buck, Randy A. Craft, Dale J. Hoff,
Susan L. Tank, and Theodore T. Buerger

We conducted radiotelemetry monitoring studies of the effects of an organophosphorus (OP) pesticide on 6 bird species from 1987 through 1991. Bird species studied included the great horned owl (*Bubo virginianus*), American kestrel (*Falco sparverius*), ring-necked pheasant (*Phasianus colchicus*), northern bobwhite (*Colinus virginianus*), American robin (*Turdus migratorius*), and blue jay (*Cyanocitta cristata*). Different radio-tracking techniques were used based on the biology and behavioral ecology of the species. Herein we discuss the telemetry equipment and methods for locational data capture, data handling, and data analysis. Locational data were georeferenced to X-Y grid coordinates to facilitate the use of spacial data analysis in relationship to study site boundaries and habitat features. Methods were employed that minimized human-induced disturbance of normal animal behavior. Telemetry relocation data were used to evaluate survival, home range use patterns, agricultural habitat use, and activity rates. Successes and problems with monitoring techniques and analytical methods are discussed.

Radio technology has advanced dramatically with the development of micro-electronics and miniature batteries, providing greater equipment portability for the scientist and smaller transmitter packages with sufficient battery life for use on small animals such as passerines. These improvements have increased the field radio monitoring of avian behavior and survival in studies to identify and quantify nontarget effects of agricultural pesticides (McEwen and Brown 1966; Brewer et al. 1988; Blus et al. 1989; Buerger et al. 1991; Kilbride et al. 1992). Such studies have shown that wildlife radiotelemetry enhances our ability to identify wildlife behavior that increases risk of chemical exposure and allows us to measure wildlife responses to such exposure. Numerous wildlife toxicology field studies have been conducted since the mid-1980s under the regulatory process of the Federal Insecticide, Fungicide, and Rodenticide Act (FIFRA). Registrants must refute the presumption of hazard to nontarget wildlife posed by the use of their agricultural products to gain registration approval from the U.S. Environmental Protection Agency (USEPA) (1982). The USEPA produced a guidance document for conducting such field studies in which the authors acknowledge radiotelemetry as a useful technique for monitoring animal survival and behavioral modifications related to pesticide applications

(Fite et al. 1988). Studies conducted under FIFRA guidelines often entail complex study designs that involve monitoring of exposure and response of several wildlife species. Radiotelemetry studies are often designed for study of specific species or groups of species that have similar biology or life histories (e.g., raptors versus upland birds). A study designed to monitor passerine use of a corn field may not incorporate the same techniques as a study designed to monitor upland bird use of the same field. Factors that can influence the choice of techniques include animal size (determines maximum transmitter weight and, therefore, longevity and strength of the signal), physiology and feeding ecology (determines transmitter mounting), home range characteristics and habitat use, maximum daily movement distance, and other behaviors (use of cavities, ground roosting, coveying, long periods of inactivity, etc).

We conducted studies in Iowa under FIFRA guidelines involving radiotelemetry monitoring to determine the effects of an OP pesticide on 6 bird species from 1987 through 1991. These birds represented 4 orders: Strigiformes (great horned owl, *Bubo virginianus*), Falconiformes (American kestrel, *Falco sparverius*), Galliformes (ring-necked pheasant, *Phasianus colchicus*, and northern bobwhite, *Colinus virginianus)*, and Passeriformes (American robin, *Turdus migratorius*, and blue jay, *Cyanocitta cristata*). In this paper we have concentrated on the equipment and methods of radiotelemetry monitoring that were most successful in our attempts to collect locational, survival, and behavioral data on the target animals and the statistical analysis methods for the resulting data. The intent of this paper is not to present a detailed summary of wildlife exposure and response to the test chemical; these results will be published elsewhere.

Study area and study sites

The 9 study sites were 160-acre corn fields and adjacent habitat located in Lucas and Wayne Counties of south-central Iowa. All study sites fell within a radius of about 16 km. The terrain was gently rolling with scattered ravines following drainages. Hardwood forest blocks were mixed among corn, soybean, and alfalfa crops with substantial area in Conservation Reserve Program (CRP) native grass plantations. Study sites were planted with corn, but all had grass waterways and hedgerows of hardwood trees and shrubs within their boundaries. Study sites were either treated at planting with Counter® 15G systemic insecticide nematicide (active ingredient terbufos) or planted without insecticide to provide reference sites. Counter® 15G was applied either infurrow or banded over the seed row and lightly incorporated.

Wildlife capture and telemetry methods

Great horned owls, American kestrels, ring-necked pheasants, northern bobwhites, American robins, and blue jays were selected as study species. These species were captured, leg-banded, and radio tagged on each study site. Monitoring techniques were designed primarily by group—raptors, upland birds, and passerines.

Great horned owls

The objectives of radio tracking great horned owls were 1) to determine their home range during the study period, 2) to compare their use of pesticide-treated versus nontreated habitats within their home range, 3) to monitor their survival, and 4) to facilitate repeated trapping for the purpose of collecting blood and fecal-urate samples to monitor exposure to OP pesticides. Survival analyses were not conducted because none of the radio-tagged owls died during the study.

Concentrating trapping effort on study site perimeters, we trapped 11 great horned owls from 5 May to 16 June in 1988, and 21 owls from 27 January to 14 July in 1989, using funnel and triggered-box traps (Petersen 1979), harnessed pigeons (Webster 1976), and a bow net (Tordoff 1954) automated with a rat-trap trigger mechanism. Traps were baited with rock doves (*Columba livia*), 2- to 4-week-old chickens, or pen-reared northern bobwhites. Owls were fitted with radio transmitters in a crisscross, backpack fashion (Smith and Gilbert 1981) with teflon ribbon as the harness material. We used a cage clip and glue to clasp the teflon strands together at the breast. Transmitter packages weighed 25 to 40 g and did not exceed 4% of the average weight of male or female great horned owls reported in the literature (Earhart and Johnson 1970). All owls captured were banded with U.S. Fish and Wildlife Service (USFWS) locking tarsus bands. Estimated life of the transmitters was 18 to 36 months, depending on the battery size used. We found that the owls easily carried the transmitter package even when we used size AA batteries.

We monitored 7 radio-tagged owls in 1988 and 15 owls in 1989 using a stacked null telemetry antenna system mounted on the roof of a pickup truck (Hegdal and Colvin 1986; Kenward 1987). The antenna system was calibrated at least once every 24 h of use. The accuracy of the system ranged from 20 to 100 m at distances ≤ 1.5 km, based on triangulation on transmitters placed at recorded locations unknown to the operators. Estimated bearing error of a similar vehicle mounted system was $\pm 1.44°$ (Balkenbush and Hallett 1988).

Road intersections within the study area were marked with reflective numbers that could be read easily at night. The numbered intersections were assigned their respective Universal Transverse Mercator (UTM) coordinates determined from U.S. Geological Survey (USGS) 1:24,000 quadrangle maps. Thus, a UTM grid was established for the entire study area. These numbered intersections were used as tracking stations. Additional tracking stations were established in other key locations as experience identified locations that were ideal for obtaining strong transmitter signals.

We located radio-tagged owls by triangulation (Mech 1983) from known positions (tracking stations) every 3 h during a 9-h telemetry session, which lasted from 2100 to 0600 h and was repeated 4 consecutive nights per week. Owls were tracked on a rotating, systematic schedule to prevent time-related sampling bias. Locations were plotted to the nearest 100-m grid UTM coordinate intersection on 1:24,000 USGS topographic maps.

When possible, 2 antenna-equipped vehicles were used simultaneously to obtain triangulations. Vehicle operators communicated via portable short-wave radios.

Cover types within ranges of great horned owls were mapped using aerial photographs (1:1000 scale) obtained from the U.S. Department of Agriculture Agricultural Stabilization Conservation Service (ASCS) in Salt Lake City, Utah, and current crop compliance slides from local ASCS offices in Wayne and Lucas Counties. Boundary lines and cover types were updated by projecting compliance slides over the aerial photos and tracing the current habitat and boundary delineations, generating a new cover map. Cover types were verified in the field.

Radio-tagged owls were monitored only after pesticide application in 1988. In 1989, each radio-tagged owl was monitored during 2 periods: a pre-pesticide application period and a postapplication period. Each owl was monitored for an equal number of days per time period. Therefore, the postapplication range of an owl was based on the same number of telemetry locations as the preapplication range.

Postapplication ranges for owls radio monitored in 1988 and pre- and postapplication ranges for individuals in 1989 were estimated using minimum convex polygon (MCP) (Mohr 1947) and 50% harmonic-mean activity area (HMAA) (Dixon and Chapman 1980) home-range models, calculated with the software program TELEM88 (Coleman and Jones 1988). We used both MCP and 50% HMAA to represent home ranges because the MCP can include outliers or infrequent movements by an owl, whereas the HMAA is an area of high use and may represent the portion of an owl's range where the greatest potential for pesticide exposure exists.

Habitats within the individual pre- and postapplication MCP and HMAA ranges were categorized as treated or nontreated areas, and sizes and percentages of these areas were calculated from updated cover maps using a digitizer interfaced with a personal computer. Comparison of use characteristics between the pre- and postapplication periods for each owl provided a control for intra-individual variation in range size over time.

The percent-use of treated areas for each owl was determined by plotting owl locations on cover maps and overlaying an additional transparent grid made up of multiple 1-ha cells. To account for edge habitat and location error, we oriented the grid so that 1 location for each owl was centered in a 1-ha grid cell. Grid cells containing a location and treated habitat were considered treated locations, and grid cells containing a location and nontreated habitat were considered nontreated locations. Percent-use represents the percent of all telemetry locations falling in treated grid cells for 1 owl.

We calculated the postapplication MCP and HMAA ranges, percentages of OP-treated areas within these ranges, and percent-use of treated areas from telemetry data collected in 1988. For owls monitored in 1989, we used the Wilcoxon signed rank test to detect changes in MCP sizes, percentages of treated areas within MCPs, and the percent-use of treated areas between the pre- and post-monitoring periods.

We tested for changes in the pre- and postapplication HMAAs by first separating owls into 2 groups. Owls containing no treated areas within the pre- or postapplication HMAAs were designated as the "untreated HMAA group," and owls containing treated areas within the HMAA were designated as the "treated HMAA group." Identification of significant differences in sizes of the HMAA between untreated and treated HMAA groups may be a sensitive measure of shifts in use of treated habitat by 1 of the groups (Anderson and Mytton 1990), even though treated areas occurred within the larger MCP range of some owls in the untreated HMAA group. Linear distances between the geometric centers of the pre- and postapplication HMAAs were also compared between the treated and untreated HMAA groups. Changes in size ratios and shifts in geometric centers between pre- and postapplication 50% HMAAs were compared between treated and untreated HMAA groups using the Wilcoxon rank sum test, performed with SAS (SAS Institute, Inc. 1985).

Independence of successive observations collected pre- and postapplication from individual owls was tested using Schoener ratios (Schoener 1981; Swihart and Slade 1985a) and calculated with the computer software program Home Range (Ackerman et al. 1988). The Schoener ratio was calculated using t^2/r^2, where t^2 represents the mean squared distance (Swihart and Slade 1985b) between successive observations, and r^2 represents the mean squared distance from the center of activity.

American kestrels

The objectives of radio monitoring American kestrels were 1) to assess survival rates of fledging kestrels during the study period, 2) to evaluate summer home range of post-fledging kestrels in relationship to studying site boundaries and potential for exposure, and 3) to determine whether OP pesticide application altered the rate at which the study sites were used by kestrels.

American kestrels were captured by hand in nest boxes. Adult kestrels were banded with USFWS-numbered leg bands, and each bird was assigned a unique identification number based on that band number. Adult birds were aged as hatching year (HY) or after hatching year (AHY) and sexed by plumage characteristics. Nestlings < 15 d of age were marked for identification using a numbered temporary plastic band (cable tie). Once they were of banding age (based on hatch dates), the nestlings were banded with a pre-assigned USFWS leg band.

Adult and juvenile American kestrels were fitted with transmitters modified into backpack mounts. Transmitters were attached using braided cotton strings that were crossed at mid-breast, tied, and glued. Cotton string rather than polyester string was used because it will biodegrade more rapidly to assure the bird will shed the transmitter within a reasonable time. Only those juvenile birds estimated to fledge before the project completion date were instrumented. When feasible, radio-tagged kestrels were relocated once every other day. Telemetry locations were determined both while walking using hand-held equipment and from a pickup truck with a roof-mounted antenna system. Locations also were determined using the truck-mounted unit in combination with 2 fixed-station antenna units. Most locations were obtained using hand-held equipment while walking because the small transmitters used were usually too weak to monitor consistently from nearby roads using the truck-mounted antenna. Personnel generally obtained hand-held locations by walking into the area near each occupied nest box and monitoring frequencies of the radio-tagged birds associated with the box. Each bird was tracked until 1 of the following determinations was made: the bird was still alive based on radio signal fluctuation, location was established by signal volume, or the bird was seen. Locations were plotted on a photocopy of a topographic map. Mapped locations were assigned UTM grid coordinates based on proximity to grid line intersections and recorded to the nearest 10 m. The truck mounted unit used for the Kestrel monitoring was the same as used for great horned owl monitoring.

Occasionally, the truck-mounted antenna unit was used in combination with 2 stacked, null telemetry antenna units mounted on wooden tripods. The tripods were positioned approximately 400 m apart at locations identifiable on the map boards and no closer than 200 m from the active nest box of the birds monitored. The truck was then parked at a numbered telemetry station on the nearest road. This station was separated from each tripod by 200 to 400 m. An observer stationed at each tripod antenna relayed signal bearings to a third observer stationed at the truck who plotted 3 simultaneous bearings. Subsequent locations were recorded as UTM coordinates, as described above. Each tripod antenna system was calibrated daily prior to locational data collection, periodically during monitoring, and again immediately after data collection ended. A beacon transmitter was placed in the field at a point accurately identifiable on the field map, and a protractor was used to measure the exact bearing from the tripod compass rose. The receiver was tuned to the beacon transmitter frequency and the antenna turned toward the beacon so the bearing of the null signal could be determined from the arrow indicator and the compass rose. This bearing was recorded as the "actual bearing." If no difference existed between the measured bearing and the actual bearing, the unit was considered calibrated. If a difference existed between the measured and actual bearing, the indicator needle was adjusted to the measured bearing when nulled on the beacon.

The harmonic-mean home-range method (Dixon and Chapman 1980) was used to evaluate the hunting range of family groups to assess kestrel use of treated study sites. Kestrel survival was analyzed using the Kaplan-Meier method (Krebs 1989; Pollock et al. 1989), and chi-square analysis was used to determine whether the rate of kestrel study-site use was independent of chemical application.

Ring-necked pheasants and northern bobwhites

The objectives of monitoring radio-tagged ring-necked pheasants and northern bobwhites were to 1) determine survival rates, 2) determine the amount of time spent on the study sites by the radio-tagged animals, 3) compare activity rates as determined by liquid mercury motion switches built into the transmitters, and 4) use the transmitters to relocate and recapture birds for blood and fecal-urate sampling.

Most bobwhites were captured with baited funnel traps placed on the ground. Pheasants and some bobwhites were spotlighted and netted at night using an all terrain vehicle (ATV) equipped with a gasoline powered, 120 v generator, overhead flood lights, and a spotlight. Ring-necked pheasants were equipped with bib-mounted, solar-powered, 2-staged transmitters that rested on the crop area of the birds' breasts with the antennas standing straight up behind their heads. Bobwhites were equipped with battery-powered, 2-stage transmitters mounted in similar fashion using plastic tubing necklaces. Braided cotton strings were attached to the bottoms of the transmitters and tied at the birds' backs behind their wings. Each antenna circled around to the back of the neck inside the neck collar and extended down the back toward the tail. Every transmitter had a built-in mercury switch that changed the pulse of the transmitter signal when the animal was moving, which helped determine whether the animal was alive and allowed for comparison of bird activity rate among treatment groups.

Radiotelemetry monitoring procedures for pheasants and bobwhites were very similar. Each study-site perimeter was measured, and a flag was placed every 20 m along all sides. On rolling terrain, it was necessary to use a scaled grid pattern over an aerial photo of the site to adjust the flagging to attain accuracy in placement. We used the flagged locations as telemetry monitoring stations and as the basis for a grid system that divided the study sites into cells that were 20 m on a side (400 m^2). This grid pattern was drawn to scale on clear plastic to overlay an aerial photo or scaled habitat map of the study site.

A 2-member team would radio track animals on each study site. Each team was responsible for 2 sites and would try to locate all birds on each site twice daily. Initially, 1 team member would walk an east-west border of the site while the other member walked a north-south side. We carried frequency-scanning receivers, headphones, and 5-element yagi antennas mounted on 6 m, telescopic aluminum poles. Each individual moved to the grid line intersection on the respective edge of the study site that was closest to perpendicular to the transmitter signal monitored where the respective grid line number was communicated via short wave radio. One team member then moved to a third side of the study site, located the grid line that was perpendicular to the transmitter signal, and reported this to the data recorder. The intersection of these 3 grid lines identified the grid cell containing the radio-tagged animal. We continued to work the perimeter of the study site until we triangulated the location of all transmitters that could be heard from the perimeter. This approach allowed us to locate animals with the least amount of observer-caused disturbance. After locating all transmitters possible from the field perimeter, we walked down the respective grid lines until we were close enough to the signal to verify the location and status of the animal. Finally, the team roamed the study site for transmitter signals not heard from the field perimeter, while concentrating on elevated positions where signals are more clearly received. When radio-tagged animals were located during the roaming process, we again positioned ourselves at or near 90° angles from the signal and moved toward it until we could verify the location and status of the animal. An aerial photo overlaid with the grid was used to identify the cell number of the location. The data recorder recorded the status as dead, alive, or unknown and recorded the type of determination as 1) absolute (visual observation) or 2) audio (determined by hearing signal strength fluctuations). Whether the animal was on or off the study site was also recorded and identified as the following:

1) visual location,
2) virtual location (2 observers were close enough to the signal to be certain of location), or
3) triangulated location (determined by triangulation along grid lines).

When transmitter signals were heard from off-site locations, we attempted to locate the bird at the same regularity as those birds located on site. Automobiles equipped with a magnetic, roof-mounted, omni-directional antenna were used to roam the roads of the study area to locate birds not found on or near study sites. Signals detected were tracked on foot using a hand-held yagi antenna, and the bird location was determined. The pulse rate from transmitters of relocated birds were monitored for 2 continuous minutes to record the total number of seconds birds were active, as indicated by the mercury switch.

Differences in pheasant and quail survival rates between treatment groups were determined using the Kaplan-Meier method (Krebs 1989; Pollock et al. 1989). Differences in bobwhite survival were also analyzed using the Survival by Proportional Hazards (SURPH) method (Smith and Skalski 1989). Differences in activity rates between treatment groups were determined using analysis of variance (ANOVA) with treatment and time (weekly periods following chemical application) as the independent variables and activity as the dependent variable (Sokal and Rohlf 1981). We used chi-square analysis to determine whether the rate of study site use by radio-tagged birds was independent of treatment.

American robins and blue jays

The objectives of radiotelemetry monitoring of robins and blue jays were 1) to determine the survivorship of both species and 2) to determine the rate at which both species utilized the study sites. Robins were captured with mist nets and blue jays with baited wire funnel traps placed on the ground and on platforms elevated about 1.25 m.

All robins and blue jays captured were banded with USFWS aluminum leg bands. Radio transmitters, weighing no more than 3% of total body weight, were backpack-mounted on both species. The transmitters were attached with braided cotton strings that were crossed at mid-breast, tied, and glued. Nontarget species caught in traps and nets were released immediately without banding.

We attempted to relocate all radio-tagged robins and blue jays 3 times a week through at least 21 d postapplication. Radiotelemetry monitoring was conducted using the procedures described above for pheasants and northern bobwhites, but activity rates were not monitored. The status of a bird (alive, dead, or unknown) and its position were determined for each relocation. When transmitter signals were received from off-site locations, we attempted to determine the status of the bird with the same regularity as for those birds located on site. We used antenna-equipped automobiles, as described above, to locate radio-tagged birds not found on or near study sites.

Approximately 2 weeks after trapping and radio tagging were completed, and as needed thereafter, we used fixed-wing aircraft (Cessna 172 or 182) to search for birds that were radio tagged but never relocated after release. The aircraft were equipped with a yagi directional antenna affixed to each wing strut and a single omni-directional antenna on the left wing strut. The 2 types of antennas were attached to separate receivers in the aircraft. The study area was searched systematically by flying crisscrossing transects. After birds were located by air, bird status was verified on the ground. We compared robin and blue jay survival rates and site-use rates between treatments, as described previously for bobwhites.

Results

Great horned owls

We radio tracked 22 great horned owls during the study. The radio transmitters provided excellent range (up to 2 km) and longevity (18 to 36 months). The backpack-mounting technique appeared to have few negative effects on the owls, as there were no mortalities among the radio-tagged individuals. The monitoring techniques provided adequate numbers of relocations to allow excellent home-range analyses. For example, we obtained 2051 nocturnal and 221 diurnal locations for 15 owls monitored in 1989. Accurate referencing of all relocations to the UTM coordinates facilitated the use of both the HMAA and the MCP home-range analysis methods. The development of computer files of the georeferenced relocation data allowed for generation of home-range boundaries overlaid on study-site boundaries. The MCP analysis provided an assessment of maximum overall home range size, the percent of the home range consisting of treated croplands, and the percent of relocations occurring on treated areas (Table 5-1). The HMAA analysis was used to relate home range to study-site boundaries (Figure 5-1) and demonstrate areas of varying concentrations of use in relation to study sites. The HMAA analysis provides 95%, 75%, 50%, and 25% use contours overlaying a study site (Figure 5-2).

Table 5-1   Minimum convex polygon home range characteristics and mean percent of great horned owl use of treated areas pre- and post-pesticide application in Iowa, 1989

	Total # locations	Size of MCP (ha)		% of MCP treated		% Owl use of treated areas	
		Pre	Post	Pre	Post	Pre	Post
Mean	76	207	200	12	26	18	22
Range	14–114	82–400	66–589	0–49	0–46	0–71	0–71

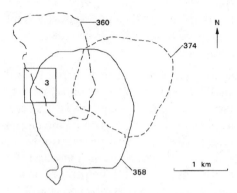

Figure 5-1 Harmonic-mean activity area of 3 great horned owls (358, 360, and 374) relative to the boundaries of study site 3 in Iowa, 1989.

Figure 5-2 95%, 75%, 50%, and 25% use contours within the harmonic-mean activity area of great horned owl number 358 relative to the boundaries of study site 3 in Iowa, 1988. The innermost contour (25%) depicts the area of most concentrated use.

The 25% use contour depicts the most intensely used portions of the home range, often referred to as the core area (Dixon and Chapman 1980). The 50% HMAA use contour was used to determine whether there were shifts in geometric centers of activity between pre- and post-treatment monitoring periods (Table 5-2).

Table 5-2 Pre- and postapplication changes in 50% HMAA size and shifts in geometric centers of activity for great horned owls using pesticide-treated areas in Iowa, 1989

Treatment Group	50% HMAA size (ha)		% Treated area in 50% HMAA		Distance between geometric centers (m)
	Pre	Post	Pre	Post	
Untreated					
Mean	63	46	–	–	464
Range	21–138	37–54	–	–	60–1660
Treated					
Mean	49	44	21	21	482
Range	3–109	24–77	0–70	0–50	60–1360

American kestrel

Eighteen AHY females and 33 male and female HY kestrels were radio tagged. Two male AHY birds were captured and radio tagged, but no relocations were obtained because of 1 known radio failure and 1 bird not being relocated after release. Therefore, all relocation data were collected on AHY females and HY birds. The kestrel transmitter signal was generally too weak to monitor the birds consistently from the roads. We obtained most relocations using hand-held yagi antennas while walking the study sites, but triangulation with the truck-mounted antennas was effective in some cases. A total of 549 kestrel relocations were obtained during the study. Twenty-five radio-tagged kestrels on untreated sites yielded 256 relocations, 62% of which were within study-site boundaries. Twenty-six radio-tagged birds on treated sites provided 293 relocations, 73% of which were within study site boundaries.

Home ranges were plotted for family groups (female and fledglings) because no males were monitored. The 95% HMAA contours were calculated and plotted (e.g., Figure 5-3). Home-range data were collected on 1 to 3 kestrel family groups inhabiting nest boxes on all but 1 study site. The HMAA home-range sizes varied from < 0.01 km^2 to 1.06 km^2. Mean home-range sizes for untreated and treated sites were 0.56 km^2 and 0.33 km^2, respectively. Most of the home-range areas were oval or circular in shape, and range overlap was common on or near sites having more than 1 occupied nest box (Figure 5-3).

A 2 × 2 chi-square test comparing the number of on-site telemetry locations to the number of off-site locations postapplication (post-planting for untreated sites) indicated that site use frequency was not related to chemical application ($P > 0.05$).

The telemetry data provided a sufficient sample size to use the Kaplan-Meier method to test for difference in survival between post-fledging HY kestrels inhabiting treated sites and those inhabiting untreated sites (Table 5-3). The survival analyses were conducted

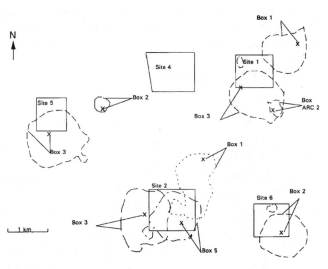

Figure 5-3   95% harmonic-mean activity area for American kestrel family groups related to nest box locations on or near pesticide study sites 1, 2, 4, 5, and 6 in Iowa, 1992.

Table 5-3   Kaplan–Meier survival rates for post-fledging, hatching year American kestrels using untreated sites compared to those using sites on which pesticide was applied in a band over the seed furrow in Iowa, 1992

Treatment	n	Survival rate[a]	Variance	95% Confidence limits
Untreated	18	1.0	0.000	1.0000–1.0000
Treated	11	0.82	0.014	0.5903–1.0461

[a] Z-score -1.56, $P = 0.06$.

using birds grouped by chemical application method (i.e., banded versus infurrow) on the study site they occupied.

Ring-necked pheasants

The capture method of spotlighting and netting was extremely time consuming, requiring 8 to 10 h per bird captured. Eighteen pheasants were captured, of which 14 were radio tagged and monitored in 1988: 7 were from each of the untreated and treated sites. Radio tagging facilitated recapture and repeated nonlethal sampling of individual birds. Pheasants captured on untreated sites yielded 95 relocations, of which 29% were within study-site boundaries. Birds captured on treated study sites yielded 164 relocations, of which 35% were within study-site boundaries.

The bib-mounted, solar-powered transmitters worked well. Transmitter signals were received easily from the study-site edges, and we had no problems with transmitter failures. Several transmitters were recovered in working condition 1 year after placement on birds. Five pheasants inhabiting untreated areas and 5 inhabiting treated areas died during the study. Survival rates were not different between the 2 groups and were independent of chemical application.

Northern bobwhite

We captured 94 northern bobwhites during the course of the study, and 70 were radio tagged: 23 on untreated sites and 47 on treated sites. The breast-mounted, battery-powered transmitters used on bobwhites were adequate; however, they provided a weaker signal than the pheasant transmitters, and we experienced about a 10% failure rate. The signal often could not be heard from the edge of the study sites, requiring us to approach the birds more closely to get a reliable location. The weaker signal was probably due to

Table 5-4   Contingency table analysis of the number of northern bobwhite relocations on and off study sites by treatment in Iowa, 1989

		Infurrow	Banded	Control	Total
On site	Observed	34.00	56.00	24.00	114
	Expected	17.11	50.53	46.35	
	Cell X^2	16.66	0.59	10.78	
Off site	Observed	93.00	319.00	320.00	732
	Expected	109.89	324.47	297.65	
	Cell X^2	2.59	0.09	1.68	
Total		127.00	375.00	344.00	846

Overall chi-square 32.40, $P < 0.001$.

the smaller battery used in these transmitters. The mercury activity switch worked well and effectively facilitated determination of the bird's status. We relocated the 70 radio-tagged bobwhite 846 times during the study. Overall, 13% of the locations were on site, and 87% were off site. A 2 × 3 contingency table analysis (Table 5-4) employing a chi-square test showed that the proportion of locations of radio-tagged bobwhites on the study sites was different among treatments. These data indicated that bobwhites captured on control sites spent less time on study sites than did bobwhites captured on either infurrow or banded treatment sites. We believe these differences were related to habitat characteristics of the study sites and not to chemical application.

Bobwhite activity rates, as determined by monitoring the signal mode of mercury-switch activity indicators within the radio transmitters, were not different among treatments. This held true for overall activity rates and for activity rates grouped into 1-week intervals. We generated survival rates by treatment group using the Kaplan-Meier method and found no differences among treatments (Tables 5-5 and 5-6).

Table 5-5   Survival rate, variance, and the 95% confidence intervals by treatment and site for radio-tagged northern bobwhite during the 1989 Iowa pesticide field study

Treatment	n	Survival rate	Variance	95% Confidence intervals
Infurrow	10	0.429	0.035	0.062–0.795
Banded	22	0.398	0.012	0.185–0.613
Control	24	0.583	0.024	0.282–0.885

Table 5-6   Comparisons of radio-tagged northern bobwhite survival rates between treatments using the Z-test to calculate probabilities in Iowa, 1989

Treatment Comparisons	Z-Score	P
Infurrow vs. banded	0.1370	0.444
Infurrow vs. control	-0.6391	0.261
Banded vs. control	-0.9773	0.163

American robins and blue jays

Eighty blue jays were radio tagged, 25 of which inhabited untreated study sites and 29 of which inhabited treated study sites. Of the blue jays on treated sites, 15 were on sites where infurrow applications were made, and 14 were on sites where banded applications were made. Fifty-four robins were radio tagged, with 27 on untreated sites and 53 on treated sites. Of those robins on treated study sites, 27 were on sites where infurrow applications were made and 26 were on sites where banded applications were made.

The small radio transmitter used on robins and blue jays produced a relatively weak signal. Both species spend considerable time on the ground, and the transmitter signals are difficult to receive when this is the case. Most of the transmitter signals were not heard from the perimeter of the study sites. We usually went directly to key locations on the study sites to receive the signal and worked toward the bird until we made positive deter-mination of location and status. The grid system was still utilized to provide coordinates for recording bird locations. Robins were more consistently on the study sites than were blue jays. However, the larger number of radio-tagged blue jays allowed for the collection of adequate data to conduct survival analyses.

A staggered entry design was necessary for survival analysis because birds were captured and radio tagged over several weeks. Again, the Kaplan-Meier method provided a satis-factory analysis, and no differences in survival were detected among groups. Examples of the output from this method are provided in Tables 5-5 and 5-6.

Discussion

We reduced variability and increased the quality of our telemetry data by strictly adher-ing to standard operating procedures. From our experience in these studies, we felt that the greatest tracking precision for some avian species could be obtained by employing stationary antenna towers. In the absence of a tower system in our study, the truck-mounted antenna system was a good alternative. Using county roads, which essentially created a 1-mile square grid system over the study area, and preselecting tracking sta-tions that were used regularly throughout the study added consistency and dependability to the data. It was important to calibrate the antenna system daily and to frequently test our triangulation accuracy by placing beacon transmitters in the field. This allowed us to estimate the degree of error in our triangulation process, which is necessary when evalu-ating the tracking system for use with different species.

The biology and ecology of the wildlife species monitored should be a determining fac-tor in the decision to use a tower- or truck-mounted antenna system. The great horned owl, for example, is large and strong enough to carry a transmitter package capable of emitting a signal detectable from long distances. This facilitated tracking the birds from greater distances using the truck-mounted antenna system. Additionally, the great horned owls had a relatively large home range. Therefore, the error polygon around each triangulated location was not large enough to represent a significant error relative to the entire home range. American kestrel home ranges were smaller than those of great horned owls, but they were large enough to facilitate a tower- or truck-mounted antenna system. However, kestrels could not carry a large transmitter. The weaker signal was not sufficiently audible from distances greater than 400 to 600 m. Therefore, we had to aug-ment the truck-mounted antenna system by placing researchers in the field with hand-held antennas. This added substantially to the variability in the method and certainly increased the error factor of triangulation. Smaller birds, such as American robins and blue jays, carried smaller, weaker transmitters and, while nesting, had home ranges that were so small that the error polygons associated with triangulation from distant truck-mounted antennas could be larger than the home range under some circumstances. This situation eliminated the advantages of the truck-mounted system, requiring us to use hand-held equipment for these species. Using semi-permanent antenna towers located around the nest greatly increased the quality and efficiency of telemetry monitoring during periods of adult foraging to feed nestlings, even though the towers needed to be spaced ≤ 400 m from the target nest.

Hand-held antennas are generally less accurate for direction determination and triangulation than multi-element antennas operated from greater distances above ground (Mech 1983; Kenward 1987). The differences in experience and hearing acuity among individuals using hand-held antennas introduces additional error and variability to the data. The presence of people walking across the study sites may seriously influence the position and behavior of target animals. This may introduce bias in studies designed to determine the amount of time a radio-tagged animal spends on a study site and could also bias estimates of home range. These potential problems with hand-held antennas must be balanced against the advantages of obtaining a visual record of an animal's position and status (i.e., dead versus alive) offered by using hand-held equipment. In a study strictly designed to monitor animal survival, researchers walking the study sites with hand-held antennas may not bias the data collected and may provide more dependable data than triangulation methods. However, the use of mercury switches and triangulation may provide satisfactory results in many cases. While we had satisfactory performance from the transmitter designs used in this study, researchers should be cautious when choosing the manufacturer and design of transmitters. Field testing different designs with the species intended for study will circumvent many problems and failures during the radiotelemetry study.

Georeferencing each location point by using a grid system that is relative to a map projection is advantageous. It immediately provides X-Y coordinates that can be entered into almost any computer database and that are compatible with most home-range analysis programs. If habitat analysis is a study objective, data referenced to a map (or aerial photo) greatly facilitates habitat assessment. Evaluation of movement patterns or habitat use may involve consideration of topographic relief. Use of UTM coordinates, which are printed on all USGS Quadrangle (quad) Maps (White and Garrott 1990), as a georeferencing grid makes the data compatible with USGS digital terrain data or quadrangle maps from which topographic data can be computer-generated and digitized onto home range maps. However, any accurate grid system can be used for home range analyses provided that its scale is known. Global positioning systems (GPS) hold promise for future radiotelemetry research. Neither a grid system nor antenna stations are required when using a GPS, which allows triangulation from any point and calculates UTM coordinates for any location on earth.

Serial telemetry data involve a time dimension, and the closeness of 2 animal locations in time influences whether the 2 locations are statistically independent from one another. Locations that are not statistically independent of one another are considered to be autocorrelated. The importance of autocorrelation has been discussed by several authors (Dunn and Gipson 1977; Dunn 1978; Schoener 1981; Swihart and Slade 1985a, 1985b; White and Garrott 1990). The decision whether to design our telemetry sampling scheme to assure statistical independence between locations was related to study objectives. The best estimate of home range is accomplished by avoiding autocorrelation in the telemetry data. However, our objectives for the owl and kestrel studies included estimates of study-site use rates, which required relatively frequent relocations. Limiting the frequency of relocations to satisfy statistical independence would have produced too little data to accurately estimate site use. We weighed the importance of avoiding autocorrelation against the objectives of the study. In both our great horned owl and American kestrel studies, the autocorrelation of the telemetry data was significant as measured by the Schoener ratio. However, since these 2 species are primarily perch-and-hunt predators, it was important to determine location at relatively close intervals so as not to miss the occurrence of predatory events relative to treated sites. We did establish set sampling periods which assured equal sampling effort by time of day and day of the week, during which there were a significant number of relocations. We are confident that the locational data we collected fully represented the home range of the birds monitored. We believe that, in studies such as these, statistical independence should be established for the sampling period; however within each sampling period, frequent locations should be collected to satisfy all study objectives regardless of autocorrelation concerns. If autocorrelation becomes a concern when estimating home range, a subsampling routine can be used to select 1 location from each sampling period for each bird and the home range can be estimated from the subset of locations (Swihart and Slade 1985b).

Delineating true reference sites is difficult in field studies in agricultural areas. Wild birds are highly mobile and can quickly move beyond the boundaries of a study site designed as a control by applying no chemicals. Therefore, mapping of all chemical applications (not limited to the test chemical) in a study area, including all study sites and adjacent land to the extent of measured animal home ranges, is important for clarification of exposure and response scenarios in study animals. Radiotelemetry monitoring is the only practical way to correlate the exposure of an animal and its location. The validity of using reference animals for comparison of such variables as survival and brain or plasma cholinesterase activities is greatly enhanced by telemetry data that show that the reference animal did not utilize any land that was treated by pesticides of the type applied to the study sites.

Conclusions

In general, the effectiveness of radiotelemetry in avian toxicology studies is primarily influenced by the biology and ecology of the test species as related to the limitations of the telemetry technology. The weight of the transmitter package relative to animal body size is one of the major limiting factors in studying small birds. Also, variability in locational data and direct disturbance of radio-tagged birds could be reduced by development of automated data recording systems that employ stationary antenna towers. However, the effectiveness of this technology is still limited by the relationship between the degree of error inherent in the system and the magnitude of the target bird's movements during the study. While radiotelemetry is well suited for monitoring short-term survival, it may be technologically limited when used to determine home range, study-site use patterns, behavior, and habitat use in some species. Studies applying telemetry technology to these areas of investigation must be designed on a case-by-case basis related to the many variables discussed in this chapter.

References

Ackerman BB, Laban FA, Garton EO, Samuel MD. 1988. User's manual for the program home range. For. Wildl. Range Exp. Stn. Tech. Rep., Univ. of Idaho, Moscow. 78 p.

Anderson DE, Mytton WR. 1990. Home-range changes in raptors exposed to increased human activity levels in southeastern Colorado. *Wildl Soc Bull* 18:134–142.

Balkenbush JA, Hallett DL. 1988. An improved vehicle-mounted telemetry system. *Wildl Soc Bull* 16:65–67.

Blus LJ, Staley CS, Henny CJ, Pendleton GW, Craig TH, Craig EH, Halford DK. 1989. Effects of organophosphorus insecticides on sage grouse in southeastern Idaho. *J Wildl Manage* 53:1139–1146.

Brewer LW, Driver CJ, Kendall RJ, Zenier C, Lacher Jr TE. 1988. Effects of methyl parathion in ducks and duck broods. *Environ Toxicol Chem* 7:375–379.

Buerger TT, Kendall RJ, Mueller BS, DeVos T, Williams BA. 1991. Effects of methyl parathion on northern bobwhite survivability. *Environ Toxicol Chem* 10:527–532.

Coleman JS, Jones III AB. 1988. User's guide to TELEM88: computer analysis system for radio-telemetry data. Dep. Fish. Wildl., Virginia Polytechnic Institute and State Univ., Blacksburg. Res. Ser. No. 1. 49 p.

Dixon KR, Chapman JA. 1980. Harmonic mean measurement of animal activity areas. *Ecology* 61:1040–1044.

Dunn JE. 1978. Optimal sampling in radiotelemetry studies of home range. In: Shugart Jr HH, editor. Time series in ecological processes. Philadelphia: SIAM. p 53–70

Dunn JE, Gipson PA. 1977. Analysis of radiotelemetry data in studies of home range. *Biometrics* 33:85–101.

Earhart CM, Johnson NK. 1970. Size dimorphism and food habits of North American owls. *Condor* 72:251–264.

Fite EC, Turner LW, Cook NJ, Stunkard C. 1988. Guidance document for conducting terrestrial field studies. Washington DC: USEPA. EPA–540/09–88–109.

Hegdal PL, Colvin BA. 1986. Radiotelemetry. In: Cooperrider AY, Boyd RJ, Stuart HR, editors. Inventory and monitoring of wildlife habitat. Denver CO: U.S. Dep. Inter., Bur. Land Manage. Serv. Cent. p 679–698.

Kenward RE. 1987. Wildlife radio tagging. New York: Academic. 222 p.

Kilbride KM, Crawford JA, Williams BA. 1992. Response of female California quail *(Callipepla californica)* to methyl parathion treatment of their home ranges during the nesting period. *Environ Toxicol Chem* 11:1337–1343.

Krebs CJ. 1989. Ecological methodology. New York: Harper and Row. 654 p.

McEwen LC, Brown RL. 1966. Acute toxicity of dieldrin and malathion to wild sharp-tailed grouse. *J Wildl Manage* 30(3):604–611.

Mech LD. 1983. Handbook of animal radio tracking. Minneapolis MN: University of Minnesota. 107 p.

Mohr CO. 1947. Table of equivalent populations of North American small mammals. *Am Midl Nat* 37:233–249.

Petersen L. 1979. Ecology of great horned owls and red-tailed hawks in southeastern Wisconsin. Dep. Nat. Resour. Tech. Bull. No. 111. Madison, WI. 63 p.

Pollock KH, Winterstein SR, Bunck CM, Curtis PD. 1989. Survival analysis in telemetry studies: the staggered entry design. *J Wildl Manage* 53(1):7–15.

SAS Institute, Inc. 1985. SAS User's Guide, Basics Version 5. Cary NC: SAS Institute, Inc. 956 p.

Schoener TW. 1981. An empirically based estimate of home range. *Theor Popul Biol* 20:281–325.

Smith DG, Gilbert R. 1981. Backpack radio transmitter attachment success in screech owls *(Otus asio)*. *North Am Bird Band* 6:142–143.

Smith SG, Skalski JR. 1989. Assessing the fate of wild populations in an epidemiological framework. In: Weigmann DL, editor. Proc. Nat. Pesticide Research Conf. Blacksburg VA: Virginia Polytechnic Institute and State Univ. p 407–421.

Sokal RR, Rohlf FJ. 1981. Biometry. San Francisco CA: Freeman. 859 p.

Swihart RK, Slade NS. 1985a. Influence of sampling interval on estimates of home-range size. *J Wildl Manage* 49:1019–1025.

Swihart RK, Slade NS. 1985b. Testing for independence of observations in animal movements. *Ecology* 66:1176–1184.

Tordoff HB. 1954. An automatic live-trap for raptorial birds. *J Wildl Manage* 18:281–284.

Webster HM. 1976. The prairie falcon: trapping the wild prairie. In: Burdett AJ, editor. North American falconry and hunting hawks. Denver CO: North American Falconry and Hunting Hawks. p 153–177.

White GC, Garrott RA. 1990. Analysis of wildlife radio-tracking data. San Diego CA: Academic. 383 p.

[USEPA] U.S. Environmental Protection Agency. 1982. Pesticide assessment guidelines, subdivisionE, hazard evaluation: wildlife and aquatic organisms. Washington DC: USEPA. EPA–540/09–82–024 (NTIS PB83—153909).

Chapter 6

Risk assessment of rodenticides through use of telemetry and other methods: 5 examples

Kathleen A. Fagerstone and Paul L. Hegdal

Note: These studies were conducted by the Denver Wildlife Research Center (DWRC) when it was part of the U.S. Department of the Interior, U.S. Fish and Wildlife Service. The DWRC was transferred to the U.S. Department of Agriculture, Animal and Plant Health Inspection Service in 1985.

Radiotelemetry is an important tool used for monitoring both the effectiveness and the nontarget hazards associated with use of rodenticides. This chapter discusses 5 large-scale field studies conducted by the DWRC over the last 15 years. Various techniques were used to assess efficacy, including direct methods for censusing populations (telemetry and visual counts) and indirect methods (active burrow counts and mound counts). Several methods were used also to assess nontarget hazards to mammals and birds, including radiotelemetry, carcass searches, sightings of marked individuals, mark-recapture trapping, and nest surveys. The use of radiotelemetry for certain species, but not others, in these studies is discussed. The discussion also focuses upon the problems associated with radiotelemetry and its relationship to risk assessment. Of critical importance to the performance of effective field trials is the ability to obtain laboratory toxicology information on potential nontarget species before proceeding to the field; this assures that the species chosen for study will be those at greatest risk.

Prior to the formation of the U.S. Environmental Protection Agency (USEPA) in 1972, many pesticides were registered without extensive data on efficacy and potential hazards to wildlife. Of particular concern to the USEPA were the acutely toxic vertebrate pesticides, which were known to be toxic to a broad spectrum of wildlife. In 1973, the USEPA planned to hold formal hearings to determine whether the uses of some rodenticides should be canceled or amended to reduce nontarget risks. However, during prior informal hearings, the USEPA determined that additional scientific information was needed on rodenticides, particularly on the efficacy to target species and the hazards to nontarget wildlife; therefore, the formal hearings were canceled. Because very little wildlife toxicology information had been submitted for the most widely used field rodenticides such as strychnine, 1080, and zinc phosphide, the USEPA solicited information on these compounds. In June 1974, an interagency agreement between the USEPA and the U.S. Fish and Wildlife Service (USFWS) through the DWRC was signed to conduct field studies with these rodenticides. The goal was to develop data on the extent of unintentional poisoning of nontarget wildlife species. Over the next decade, the DWRC conducted a series of field studies, using USEPA and USFWS funds, designed to obtain information about target efficacy and nontarget mortality during operational use of strychnine, 1080 (sodium monofluoroacetate), and zinc phosphide rodenticides. Evaluation of these studies and their methodology formed the basis for the field study guidelines developed in 1988 by the USEPA.

The 1988 Terrestrial Field Study guidelines (Fite et al. 1988) include 2 types of field studies for wildlife—screening and definitive. The screening study grossly determines effect versus no effect. In contrast, the definitive study quantifies the magnitude of effects previously identified. The studies conducted by the DWRC and discussed in this chapter are primarily screening studies because they assess the potential for acute toxic effects (such as direct poisoning and death) and immediate sublethal toxic effects that might affect behavior and survival and do not necessarily quantify risk. Although risks are quantified for some wildlife species, these studies do not fit the guideline definition of "definitive studies" (Fite et al. 1988) for 2 reasons. First, definitive studies are designed as manipulative experiments with control and treated plots to allow quantification of the magnitude of mortality, reproductive effects, and long-term survival; in contrast, most of the DWRC studies were designed as observational studies to assess acute mortality to wildlife species following an operational treatment of a rodenticide. Second, definitive studies monitor treatment effects on only one or a few species that are already believed to be affected; the DWRC studies were designed to monitor a number of species to determine those at risk from the pesticide application. In addition, the studies were conducted at only 1 site, rather than at 8 to 14 sites, as recommended by the USEPA guidelines.

In this chapter, we describe 5 large-scale rodenticide field studies conducted by the DWRC since 1975: 1) hazards to seed-eating birds and other wildlife associated with surface strychnine baiting for Richardson's ground squirrels *(Spermophilus richardsonii)*, 2) and 3) hazards to wildlife associated with underground strychnine baiting for pocket gophers *(Thomomys talpoides)* using a burrow builder versus handbaiting, 4) hazards to pheasants *(Phasianus colchicus)* and cottontail rabbits *(Sylvilagus floridanus)* associated with zinc phosphide baiting for microtine rodents *(Microtus* spp.) in orchards, and 5) hazards to wildlife associated with 1080 baiting for California ground squirrels *(Spermophilus beecheyi)*. Techniques used to assess rodenticide efficacy against the target species included direct methods for censusing populations (telemetry and visual counts) and indirect methods (active burrow counts and mound counts). Methods used to assess nontarget hazards to mammals and birds included telemetry, carcass searches, sightings of marked individuals, mark-recapture trapping, and nest surveys. This chapter emphasizes telemetry as an important tool for monitoring both the effectiveness and nontarget hazards associated with use of rodenticides.

All radio transmitters used in the studies (except those used on chipmunks in Study 3) were designed and built by the DWRC Electronics Laboratory in the 164 MHZ band.

Radio-equipped animals were followed using mobile radio-tracking vehicles equipped with roof-mounted, dual yagi antennas and voice radio communication systems (Hegdal and Gatz 1978; Hegdal and Colvin 1986). The antennas could be rotated from inside the vehicle, and radio bearings were indicated on a 360° protractor by a pointer attached to the antenna mast. Coaxial cables from the antenna were attached to a hybrid junction box, which allowed switching from in-phase to out-of-phase (null) operation. We used Model LA12 receivers (built by AVM Instrument Co., Champaign, Illinois) for all radio tracking. Hand-held loop and yagi antennas were employed for portable field use. Animal locations were determined in 1 of 4 ways: 1) hand tracking by walking out the radio-equipped animal using a receiver and hand-held antenna, 2) vehicle tracking using 2 vehicles and obtaining simultaneous bearings from 2 vehicles parked at previously mapped locations, 3) vehicle tracking using 1 vehicle by plotting a bearing to the animal and a second bearing to a beacon transmitter, and 4) aerial searches for missing or lost animals.

Hazards to seed-eating birds and other wildlife associated with surface strychnine baiting for Richardson's ground squirrels

Strychnine is an acutely toxic rodenticide with LD50s to mammals of 0.7 to 27 mg/kg and to birds of 2.9 to 161 mg/kg (LaVoie unpublished data). The DWRC evaluated wildlife hazards associated with surface baiting with strychnine for Richardson's ground squirrels on rangeland in south-central Wyoming (Hegdal unpublished data). In this area, cattle ranching and farming are the principal agricultural activities; ground squirrels compete with livestock for forage and can cause considerable damage to agricultural crops adjacent to rangeland. During late April and early May 1976, approximately 3650 ha were treated with 0.5% strychnine-treated oats to reduce damage caused by ground squirrels. Approximately 1 tablespoon of bait was placed at or near each ground squirrel burrow. Application rate varied between 1 and 2 kg/ha. Treated areas were scattered between Saratoga and Encampment, Wyoming.

The reduction in ground squirrel populations was measured on 6 randomly selected plots by plugging ground squirrel burrows then recording the number reopened after 48 h; the procedure was conducted immediately before treatment, 3 d after treatment, and 1 month after treatment. Effectiveness varied, with some plots showing almost no reduction in ground squirrel populations while others indicated up to 85% control ($\bar{x} = 2.8\%$). Chi-square analysis showed a significant change in populations after treatment on all but 2 plots. During carcass searches on 19.5 ha, an average of 3.5 dead ground squirrels/ha were located above ground; carcasses were left undisturbed to allow assessment of secondary risks of strychnine poisoning to mammalian predators and raptors. Residue analyses were conducted to confirm cause of death for predators and raptors.

Nontarget small mammal populations were censused by live trapping before and after treatment. Because populations were low and numbers trapped were probably inadequate to detect significant treatment effects, no conclusions can be drawn about strychnine effects on small mammals.

Previous studies had shown evidence of hazards to some seed-eating birds at strychnine baiting sites (Rudd and Genelly 1956; Hegdal and Gatz 1976), so an extensive literature search was conducted to determine focal species. Based on this search, we evaluated risk to mourning doves *(Zenaida macroura)* by radiotelemetry, to horned larks *(Eremophila alpestris)* by marking, and to other seed-eating birds by carcass searches conducted intensively on 19.5 ha for 14 d after treatment. Mortality of mourning doves was assessed by radiotelemetry because these birds can move great distances and would be difficult to census by mark-recapture techniques. We captured 84 mourning doves using Kniffin collapsible bait traps (Reeves et al. 1968) and 3-cell Potter traps. We attached colored and numbered leg streamers to birds to facilitate locating individuals. We also attached a radio transmitter by gluing the transmitter to either a latex rubber harness or a 3 × 3-cm latex rubber patch that was then glued to clipped feathers on the dove's back. The latex degraded in sunlight, and transmitters dropped off birds after 30 d to 3½ months. Most transmitters quit functioning after 3 to 4 weeks, but a few functioned for up to 3 months. Of the 84 radio-equipped doves, 59 were followed through treatment: 36 (61%) of these survived the treatment or died of causes unrelated to treatment (16 survived, while 20 either were killed by predators, died in untreated areas and did not use treated areas, or contained no strychnine residues); 23 (39%) died in or near treated areas and/or contained strychnine residues (Table 6-1), providing a quantitative estimate of risk. Dead doves were also located during carcass searches up to 80 d after treatment at an average of 1.1 dead doves/ha searched. In addition to structured carcass searches, we sometimes found dead doves while following radio-equipped birds; in one instance, 40 dead doves were found along one-half mile of fence; those analyzed contained strychnine residues (Table 6-2). The carcass search information could not be used to quantify risk because pretreatment densities of doves were not available.

Horned larks using areas near ground squirrel burrows were considered to be at risk; we therefore trapped 19 male horned larks on their territories using 3-cell Potter traps with a decoy horned lark in the center cell. Horned larks were too small to carry currently available transmitters for any length of time; therefore, they were marked with leg streamers. Because they are territorial during the breeding season, censuses on territories were made before and after treatment, and carcass searches were conducted on horned lark territories after treatment to quantify risk. As an additional measure of risk, we compared the number of larks trapped in control and treated areas after treatment. Populations of horned larks on treated areas were significantly affected. Of the 19 marked horned larks, 6 disappeared before treatment. An additional 3 were not observed after treatment, and their fate was unknown. Six of the remaining 10 (60%) marked territorial horned larks were killed by the bait (verified by residue analysis) between 1 and 18 d after treatment. During carcass searches on 19.5 ha, we found an average of 2.5 dead horned larks/ha. Pretreatment densities were estimated at 2.5 to 3.0 larks per ha; almost all of

Table 6-1   Summary of results from strychnine residue analysis of 27 radio-equipped mourning doves found dead in and near treated areas. Mourning doves did not arrive in Wyoming until May, after the baiting

# days tracked	Tracked in treated area	Dead # days posttreatment	Cause of death	Strychnine residue ppm	Part analyzed[a]
1	No	17	Bait	11.2	GI
1	Yes	21	Bait	146	GI
1	No	22	Unknown	NLT 0.5[b]	GI
2	Yes	17	Bait	19	GI
4	No	14	Bait	96	GI
4	Yes	15	Bait	51	GI
5	No	46	Probable bait kill	B	B
6	Yes	64	Bait	8.2	GI
7	Yes	55	Probable bait kill	B	B
8	Yes	43	Possible predator kill	B	B
9	Yes	29	Unknown	NLT 0.5	GI
9	Yes	55	Bait	3.9	IO
10	Yes	54	Bait	5.6	GI
11	Yes	21	Unknown	NLT 0.5	GI
15	Once	36	Bait	25	GI
16	Yes	51	Bait	<0.5[c]	WB
20	Yes	35	Bait	5.5	GI
21	Yes	36	Probable bait kill	B	B
22	Yes	33	Bait	62	GI
22	Yes	71	Probable bait kill	B	B
22	Yes	85	Probable bait kill	B	B
27	Yes	48	Bait	8.6	GI
30	Yes	50	Bait	0.96	GI
32	Yes	60	Bait	<0.5	WB
34	Once	47	Probable bait kill	B	B
36	Yes	58	Bait	1.2	WB
51	Yes	101	Probable bait kill	B	B

[a] Parts analyzed are as follows: GI – Gastrointestinal tract; IO – Internal organs; WB – Whole body;
[b] NLT 0.5 – Not found, if present less than 0.5 ppm
[c] < 0.5 – Found, but less than 0.5 ppm

the lark population was killed on treated areas. One bird was found dead 71 d after treatment. Table 6-2 contains strychnine residue levels for dead horned larks. On control areas, we trapped a mean of 10.5 birds per 100 trap-hours versus 0.87 per 100 trap-hours on treated areas, a highly significant difference (chi-square, $P < 0.005$). The risk to local horned lark populations in baited areas is therefore great.

Populations of vesper sparrows (*Pooecetes gramineus*) were not as seriously affected. Two of 16 (12.5%) vesper sparrows marked with leg streamers before treatment were found dead 2 and 4 d after treatment and contained strychnine residues. Four vesper sparrows were found during carcass searches (0.2 birds/ha), and 2 contained strychnine residues;

Table 6-2   Summary of results from strychnine residue analysis of nontarget animals found dead in treated areas

Species	# days posttreatment	Strychnine residue ppm	Part analyzed[a]
Mourning dove	2	10.3	GI
Mourning dove	2	39	"
Mourning dove	4	4.6	"
Mourning dove	7	38	"
Mourning dove	7	48	"
Mourning dove	8	13	"
Mourning dove	8	49	'
Mourning dove	11	36	"
Mourning dove	18	NLT 0.5[b]	"
Mourning dove	18	1.1	"
Mourning dove	18	3.6	"
Mourning dove	18	31	Crop
Mourning dove	18	43	GI
Mourning dove	28	4.3	"
Mourning dove	37	17	"
Mourning dove	42	1.1	"
Mourning dove	55	NLT 0.5	IO
Mourning dove	55	1.0	GI
Mourning dove	55	6.4	"
Mourning dove	56	2.0	"
Mourning dove	56	3.9	"
Mourning dove	56	34	"
Mourning dove	59	8.0	"
Mourning dove	60	0.73	"
Mourning dove	60	17	"
Mourning dove	66	6.5	"
Mourning dove	80	82	"
Horned lark	1	5.7	GI
Horned lark	1	9.0	"
Horned lark	1	9.8	"
Horned lark	2	7.5	"
Horned lark	2	8.7	"
Horned lark	2	8.8	"
Horned lark	2	12	"
Horned lark nestling	2	24	"
Horned lark	3	4.2	"
Horned lark	4	5.9	"
Horned lark	18	2.1	"
Horned lark	33	NLT 0.5	"
Horned lark	59	7.5	'
Horned lark	59	8.7	IO

Table 6-2   continued

Species	# days posttreatment	Strychnine residue ppm	Part analyzed[a]
Vesper sparrow	2	NLT 0.5	GI
Vesper sparrow	4	7.1	"
Vesper sparrow	4	39	"
Red-winged blackbird	2	6.4	GI
Red-winged blackbird	3	4.3	"
Red-winged blackbird	3	7.1	"
Red-winged blackbird	3	19	"
Red-winged blackbird	5	5.1	"
Brewer's blackbird	1	31	GI
Brewer's blackbird	2	0.67	"
Brewer's blackbird	4	52	"
Brewer's blackbird	67	0.7	IO
Yellow-headed blackbird	0	5.9	GI
Yellow-headed blackbird	1	0.89	"
Yellow-headed blackbird	4	42	'
Brown-headed cowbird	7	9.5	GI
Brown-headed cowbird	67	4.4	IO
Common crow	2	1.3	GI
Common crow	80	1.0	IO
Starling	4	14	IO
Savannah sparrow	20	43	GI
Meadowlark	10	23	GI
Meadowlark	21	2.0	WB
Meadowlark nestling	38	NLT 0.5	GI
Meadowlark nestling	38	"	"
Meadowlark nestling	38	"	"
Meadowlark nestling	38	"	"
Meadowlark nestling	38	"	"
Mallard	1	0.68	GI

[a] Parts analyzed are as follows: GI – Gastrointestinal tract; IO – Internal organs
[b] LT 0.5 – Not found, if present less than 0.5 ppm.

however, a marked reduction in vesper sparrow populations was not immediately observed.

Blackbirds were killed early in the season while migrating through the study area. During carcass searches we located 18 (0.9/ha) red-winged blackbirds *(Agelaius phoeniceus)*, 21 (1.1/ha) Brewer's blackbirds *(Euphagus cyanocephalus)*, 6 (0.3/ha) yellow-headed blackbirds *(Xanthocephalus xanthocephalus)*, 2 (0.1/ha) brown-headed cowbirds *(Molothrus ater)*, 1 (0.05/ha) common crow *(Corvus brachyrhynchos)*, 1 starling *(Sturnus vulgaris)*, 1 Savannah sparrow *(Passerculus sandwichensis)*, and 1 western meadowlark *(Sturnella neglecta)*. Those that could be tested all contained strychnine residues (Table 6-2). During radio tracking and other activities, we also found the following carcasses that contained strychnine residues: 4 Brewer's blackbirds, 5 red-winged blackbirds, and 1 mallard *(Anas platyrhynchos)*.

Previous literature indicates that strychnine does not cause significant secondary poisoning (Tucker and Crabtree 1970; Hegdal and Gatz 1978); however, because a potential hazard could exist for species with a low LD50 (Wood 1965; Tucker and Crabtree 1970. Consequently, we monitored secondary hazards to mammalian predators and raptors. Radiotelemetry is the only reliable method for relocating these highly mobile predators. Therefore, we captured 2 great horned owls *(Bubo virginianus)* and 1 badger *(Taxidea taxus)* that used treated areas and radio equipped them with mortality transmitters that emitted a faster signal when the animal had failed to move for a few hours. We did not detect any detrimental effects. The radio-equipped great horned owls used treated areas, fed on treated doves, and were alive 4 months after treatment. Other raptors were observed feeding on ground squirrels in the study area and were probably exposed to some level of strychnine; however, 8 active raptor nests monitored through treatment fledged young. We concluded that the strychnine bait posed minimal risk to predators.

Radiotelemetry in this study was invaluable for determining risks to seed-eating bird species, raptors, and mammalian predators. These animals moved widely and could not easily be monitored by other means. For territorial bird species, visual surveys and carcass searches on territories proved both feasible and economical. This study demonstrated minimal risk from above ground strychnine baiting to species other than seed-eating birds.

Hazards to wildlife associated with underground strychnine baiting for pocket gophers

The following 2 studies, also conducted with strychnine alkaloid grain baits, illustrate the influence that use patterns can have in determining whether nontarget risks will occur. While the above ground baiting for ground squirrels used in the previous study produced significant seed-eating bird mortality, the below ground baiting used in the following 2 studies caused no significant risk to seed-eating birds.

Burrow builder application

Throughout the United States, pocket gophers cause extensive damage to agricultural crops by clipping vegetation and mound building (Luce and Case 1981). Because of this damage, below-ground baiting with strychnine to control pocket gopher populations is frequently employed. During spring and summer 1975, the DWRC evaluated the hazards to wildlife associated with strychnine baiting for plains pocket gophers *(Geomys bursarius)* (Rudd and Genelly 1956) using a burrow builder, which makes an artificial underground burrow into which the bait is deposited. The Sherburne National Wildlife Refuge in Minnesota was divided roughly in half to form control and treated areas, and we applied 0.5% strychnine-treated bait to pocket gopher occupied habitat on 662 ha of the treated area.

Strychnine was effective in controlling pocket gophers—activity plots using a closed-hole index (Hansen and Ward 1966) showed 88% reduction in activity. Populations of other small rodents, as shown by live trapping, were quite low pretreatment, but declined even further on the treated area ($P < 0.10$), yet significantly increased on the control area ($P < 0.001$), indicating some risk to nontarget rodents.

With the underground placement of bait, we expected primary hazards to seed-eating birds to be low. However, because small amounts of bait were occasionally available to birds through inadvertent spillage as the burrow builder tube was raised or lowered into the ground, we monitored red-winged blackbirds as a focal species for seed-eating birds. This species was chosen because it is territorial in cattail marshes during the breeding season and can be successfully monitored without the use of expensive transmitters. We trapped and marked 100 territorial males on both the treated and control areas and monitored their presence through the treatment. Although some treated grain was available on the surface and marked birds were observed feeding in treated fields, detrimental effects were not detected on red-winged blackbird populations. The number of territorial male red-winged blackbirds maintaining territories was slightly higher on the treated area ($P < 0.005$) compared to the control.

Pocket gophers are an important part of the diet of raptors during some periods of the year and can also form a high percentage of the diet of some mammalian predators. Therefore, secondary risk to these predators was studied. Because telemetry is the only reliable technique to monitor movement and mortality of highly mobile predators, we equipped 36 raptors and 36 mammalian predators with radio transmitters to measure potential secondary poisoning effects. Mortality transmitters were used on mammalian predators, which could accommodate the additional weight. Many animals left the area before treatment, had transmitters fail, or did not use treated areas. On the control area, 2 red-tailed hawks *(Buteo jamaicensis)*, 1 American kestrel *(Falco sparverius)*, 1 great horned owl, 5 badgers, 5 striped skunks *(Mephitis mephitis)*, and 1 red fox *(Vulpes fulva)* were tracked daily and survived for at least 3 weeks after the treatment period. On the treated area, 2 striped skunks, 3 badgers, 2 red foxes, and 1 coyote *(Canis latrans)* frequently used the treated areas; all survived at least 3 weeks after treatment. Use of radiotelemetry in this study allowed us to follow individuals for long periods of time: the transmitters functioned for 37 to 75 d on raptors, 89 to 132 d for skunks, and 77 to 143 d for larger mammalian predators. Radiotelemetry allowed us to conclude that secondary poisoning risks from underground baiting with strychnine to raptors and mammalian predators are nonexistent or low.

Hand-baiting application

Damage to conifer seedlings by pocket gophers is a major factor limiting reforestation on over 120,000 ha of forest land in the western United States (Northwest Forest Pocket Gopher Committee 1976). In conjunction with reforestation programs, the U.S. Forest Service (USFS) annually treats thousands of hectares with strychnine alkaloid-treated grain to control pocket gopher populations before tree planting. This has been shown to reduce pocket gopher damage to seedlings (Barnes 1974; Crouch and Frank 1979).

Because of USFS concerns that the underground application of strychnine bait could pose potential hazards to other species, we monitored an operational USFS baiting program conducted in 1979 on the Targhee National Forest, Idaho (Fagerstone et al. 1980). Our objectives were to assess the primary risks to nontarget small mammals (Fagerstone et al. 1980) and secondary risks to grizzly bears *(Ursus arctos horribilis)* (Barnes et al. 1985) resulting from hand baiting in underground pocket gopher burrow systems with strychnine alkaloid bait.

As in the previous study, live trapping was used to census small mammal populations, primarily deermice *(Peromyscus* spp.) on control and treated areas before and after baiting (Fagerstone et al. 1980); no significant differences occurred between pre- and post-treatment population estimates or between treated and untreated plots. Populations of chipmunks were too widely dispersed to monitor by means of trapping, so we used radiotelemetry to assess the hazards of strychnine baiting to yellow pine chipmunks *(Eutamias amoenus)*. Twenty-four chipmunks and 1 flying squirrel *(Glaucomys sabrinus)* were radio equipped and their movements monitored daily on treated areas throughout treatment using hand-held yagis and receivers. Transmitters (SM1 and SM1 mouse style with a collar attachment) were built by AVM Instrument Co., weighed between 2.5 and 5.8 g, and had a battery life of 2 to 4 weeks. Three of the 24 chipmunks died after treatment; all were scavenged by predators, and 2 contained low levels of strychnine residues (0.29 and 0.35 ppm), indicating potential risk from the strychnine. Assuming strychnine contributed to the 2 deaths, the 4.2% mortality rate did not have a significant effect on the chipmunk population.

To investigate potential hazards of secondary poisoning to grizzly bears associated with feeding on dead pocket gophers, we radio equipped 82 pocket gophers and determined their fate; we also located nests and food caches. Sixty-two pocket gophers were successfully monitored through treatment, 40 (64.5%) of which died as a result of the strychnine baiting. We excavated carcasses of 40 radio-equipped and 5 unmarked pocket gophers, located a mean of 48 cm below ground (range = 10 to 152 cm). Most carcasses were located in nests, which had a mean depth of 57 cm. We concluded that strychnine-poisoned gophers presented a negligible risk to grizzly bears because pocket gophers died below ground, usually separate from each other, and carcasses contained only small amounts of strychnine ($\overline{X} = 0.16$ mg).

In this study, risk assessment was greatly enhanced by use of radiotelemetry. Assessing risks to the low density, mobile population of chipmunks was possible because of radiotelemetry, even though numbers monitored were low. Use of radiotelemetry provided data on where pocket gophers died, levels of strychnine residues in poisoned animals, location and quantity of bait stored by pocket gophers, and levels of strychnine on recovered bait, allowing a good assessment of risks to grizzly bears. Without radiotelemetry, finding dead pocket gophers, nests, and food caches would have been possible only by excavating entire burrow systems.

Hazards to pheasants and cottontail rabbits associated with zinc phosphide baiting for microtine rodents in orchards

Throughout the United States, microtine rodents cause extensive overwintering damage to orchards by girdling and killing trees. Zinc phosphide is used extensively to control microtine rodent populations and reduce damage. Zinc phosphide is an inorganic acute rodenticide that reacts in the gastrointestinal tract of poisoned animals to form the toxic gas phosphine. The LD50 values of zinc phosphide for mammals range from 5.6 to 93 mg/kg and for birds range from 7.5 to 67.4 mg/kg (Johnson and Fagerstone 1994).

DWRC scientists evaluated the hazards associated with surface zinc phosphide baiting for controlling microtine rodents, the meadow vole *(Microtus pennsylvanicus)* and the prairie vole *(M. ochrogaster)*, in orchards in southwestern Michigan (Hegdal unpublished data). During late October and early November 1975, landowners treated about 385 ha of orchards with 2% zinc phosphide bait at 5.6 and 11 kg/ha using aerial and ground-broadcast treatments. Application rates varied between 11 kg/ha for aerial treatments and 5.6 to 11 kg/ha using ground broadcasting. The efficacy of treatment on voles was measured by randomly selecting 5 study plots each in treated and control apple orchards and trapping before and after treatment. Only 1 of the plots had a high enough vole population to make the population reduction estimate statistically valid; percent reduction on this plot was 88.5%.

Two nontarget species were thought to be at greatest risk from the treatment and were chosen for intensive study. Cottontail rabbits were thought to be at risk based on potential exposure to the cracked corn bait, and pheasants had low LD50s of 8.8 to 26.7 mg/kg (Johnson and Fagerstone 1994). Radiotelemetry was used to assess risk to both species because they are difficult to observe or census in other ways. We were able to radio equip only 8 cottontail rabbits because other available food made trapping difficult. Four of those rabbits used treated areas, survived the treatment, were tracked for 16 to 40 d, then were collected and analyzed for residues; 1 of the 4 contained low zinc phosphide residues (0.03 ppm). During carcass searches on 272 ha, we found 8 dead cottontail rabbits (0.03/ha); the 4 that could be analyzed for zinc phosphide contained residues of 0.28 to 6.2 ppm in the gastrointestinal tracts. In addition, 3 of 5 cottontail rabbits shot in and near treated areas after treatment were positive for zinc phosphide residues (< 0.01 to 0.38 ppm). These data suggest cottontail rabbits were consuming sublethal levels of bait. However, 34 live rabbits were observed during carcass searches. Because so few rabbits were fitted with transmitters and rabbit densities before and after treatment could not be estimated, the effects of zinc phosphide on rabbit populations could not be quantified. However, populations did not appear to be seriously affected.

Using night spotlighting, we captured and radio equipped 25 pheasants with a backpack transmitter. Pheasants were tracked for up to 42 d using both hand-held and truck-mounted antennas; only 1 pheasant was killed by the zinc phosphide bait. Another 4 were killed by predators, 7 were shot by hunters, contact was lost with 3 (one of which survived for 13 months until it was shot by a hunter), and 6 did not use treated areas after treatment; none of these birds analyzed contained zinc phosphide residues. Five radio-equipped pheasants were tracked in treated areas after treatment. One was found dead 3 d posttreatment and was probably killed by the bait because it contained 2.0 ppm zinc phosphide in the gastrointestinal tract; the other birds were collected after treatment and contained no trace of zinc phosphide, nor did 9 other pheasants collected by shooting after treatment. No dead pheasants were found during carcass searches, although 14 live pheasants were observed. Although risks to pheasants were not quantified, they appeared minimal.

Carcass searches were conducted on 272 ha between 1 and 14 d after treatment to assess risk to species not radio equipped. Six deermice *(Peromyscus maniculatus)* were found dead or dying and contained residues of 0.24 to 69 ppm. One blue jay *(Cyanocitta cristata)* contained 0.83 ppm zinc phosphide. Of special interest is the lack of mortality in northern bobwhite *(Colinus virginianus)*, which are very sensitive to zinc phosphide (LD50 = 12.9 mg/kg) (Johnson and Fagerstone 1994). During carcass searches 37 flocks of quail were observed, but no quail were found dead; 2 of 4 apparently healthy live birds captured after treatment contained residues of 0.06 and 11 ppm, indicating that sublethal exposure to the bait had occurred. None of 4 mourning doves collected after treatment contained residues. Although it is clear that exposure can occur, risk to seed-eating and gallinaceous bird populations from orchard baiting with zinc phosphide appears to be minimal.

We concluded that zinc phosphide-treated bait can effectively be used in orchards with relatively low risk. A few small mammals, cottontail rabbits, pheasants, and perhaps other birds are likely to be killed, but populations will not be significantly reduced.

This study relied on a variety of techniques to assess risk. Telemetry was very successful for pheasants, but less so for cottontail rabbits, which could not be captured easily. Researchers were forced to rely on more conventional techniques such as carcass searches to provide a risk assessment.

Hazards to wildlife associated with 1080 baiting for California ground squirrels

Of the 5 studies discussed in this chapter, the following study was the most comprehensive in terms of the variety of species monitored and the magnitude of telemetry use. The objectives of this study were to evaluate the primary and secondary risks to nontarget wildlife of an operational program using 1080-treated grain bait to control California ground squirrels. Compound 1080 is an acutely toxic rodenticide with LD50s to mammals of 0.1 to 60 mg/kg and birds of 2.0 to 20 mg/kg (Azert unpublished data).

The study was conducted in the foothills of the Sierra Nevada in Tulare County, California. In June 1977, 0.075% 1080-treated grain was applied by aerial application on about 25,000 ha of rangeland to control ground squirrel populations (Hegdal et al. 1986); the entire area was considered the treatment area. Four methods were used to evaluate the efficacy of 1080 grain bait on California ground squirrels (Hegdal et al. 1986): closed-hole, marked population survival, total population survival, and radiotelemetry. On 5 treated plots, mean ground squirrel reduction was 85%: 64.6% using closed-hole activity indices, 92.2% using marked ground squirrel survival, 84.1% using total ground squirrel survival, and 90% using radiotelemetry. Radiotelemetry and marked population survival probably provided the most accurate estimates. After treatment, 38 ground squirrels (2.8/ha) were found dead on the surface of the 5 ground squirrel plots. During off-plot carcass searches, 294 ground squirrels were found dead on the surface of 381.1 ha searched (0.8/ha), indicating that secondary exposure to raptors and mammalian predators could occur.

Radiotelemetry was the main technique used to assess risk to nontarget birds and mammals by monitoring movement and mortality of radio-equipped animals in and around treated areas. The radio-tracking system employed a beacon transmitter, numbered tracking stations, and mobile radio-tracking vehicles equipped with dual yagi antennas and voice radio communications systems. Animal locations were determined by triangulation. Motion sensitive mortality transmitters were used to detect deaths of larger predators. Mortality of other radio-equipped animals was detected by lack of movement of the transmitter signal and obtaining visual sightings of animals.

Mourning doves and California quail *(Callipepla californica)* were selected as representative species for assessment of risks to seed-eating birds. Twenty-one mourning doves and 3 California quail were equipped with transmitters. Although doves and quail visited treated areas and some were observed picking up 1080-treated bait, all survived the baiting. Doves readily consume oats (Hegdal unpublished data) and dove LD50 values are low enough (8.6 to 14.6 mg/kg) to cause mortality (S. P. Atzert, unpublished review of sodium monofluoroacetate—its properties, toxicology, and use in predator and rodent control, U.S. Department of the Interior, U.S. Fish and Wildlife Service), so either 1080 acts as a repellent to seed-eating birds or as an emetic. Only 1 seed-eating bird carcass (Brewer's blackbird) was found that was considered to be a treatment-related mortality, and it contained 1080 residues (0.2 ppm). Therefore, risks to seed-eating birds appeared minimal.

As in the studies previously discussed, secondary hazards to mammalian predators were evaluated using radiotelemetry. Five of 6 radio-equipped coyotes and 3 of 9 radio-equipped bobcats *(Lynx rufus)* died after treatment. Other radio-equipped mammalian predators (2 raccoons, 3 badgers, and 3 striped skunks) survived treatment, although 3 skunks were found dead during carcass searches, indicating a potential risk to that species. In addition, a domestic cat and dog died near treated areas. Although analytical methods were not available for measuring secondary toxicity when this study was conducted, the observed mortality indicated that mammalian predators, particularly canids, were very susceptible to secondary poisoning.

In contrast to the mammalian secondary mortalities, no treatment-related mortalities were observed while monitoring 9 radio-equipped raptors and carrion-eating birds, including 3 red-tailed hawks, 1 golden eagle *(Aquila chrysaetos)*, 1 great horned owl, 2 turkey vultures *(Cathartes aura)*, and 2 common ravens *(Corvus corax)*, or while monitoring 58 active raptor nests. Despite exposure to poisoned ground squirrels on the surface of the ground, and dead ground squirrels found in nests, raptors were not affected by 1080.

Carcass searches were conducted beginning the day of treatment and continuing for 2 weeks after treatment to assess mortality to small mammals and to determine whether any unanticipated nontarget animals might be susceptible to 1080 poisoning. During carcass searches on 381 ha, we found 15 Heermann's kangaroo rats *(Dipodomys heermanni)*, 8 little pocket mice *(Perognathus longimembris)*, 4 desert woodrats *(Neotoma lepida)*, 4 deermice, and 1 western harvest mouse *(Reithrodontomys megalotis)*. Residues in carcasses ranged from nondetectable to 76 ppm. Eleven desert cottontails *(Sylvilagus audubonii)* were found during carcass searches (residues ranges from nondetectable to 20 ppm), and 55 live rabbits were observed. The carcass searches showed that the 1080 bait presented primary poisoning risks to nontarget rodents and rabbits but did not allow for quantification of those risks because densities of animals were not known. The data do indicate that 1080 should not be applied in areas where there are endangered rodents and lagomorphs of concern.

An unexpected finding was some risk to insectivorous birds—1 acorn woodpecker *(Melanerpes formicivorus)*, 2 white-breasted nuthatches *(Sitta carolinensis)*, and 1 ash-throated flycatcher *(Myiarchus cinerascens)* were found during carcass searches (1080 residues ranged from nondetectable to 4.4 ppm). The birds had apparently been feeding on poisoned ants which contained 1080 residues.

Advantages and problems of using radiotelemetry for nontarget hazards assessment

The 5 studies discussed in this chapter illustrate many of the advantages and problems inherent with the use of radiotelemetry to monitor nontarget hazards. Advantages of radiotelemetry were many:

1) We were able to assess movement and mortality of animals that were highly mobile, secretive in nature or nocturnal, and for which adequate census means do not exist. These included large seed-eating bird species, raptors, mammalian predators, and nocturnal or burrowing mammals.

2) Telemetry allowed us to assess exposure by determining if and how much instrumented animals actually used treated areas.

3) Telemetry allowed recovery of carcasses for residue analysis to confirm cause of death.

4) Use of radiotelemetry can sometimes better define risk than laboratory LD50 data. For example, LD50 values for raptors with 1080 were generally low (about 10 mg/kg) and some risk was anticipated, yet raptors proved to be relatively immune to 1080 poisoning in field situations, possibly because they regurgitated poisoned tissue.

5) Home range and movement information developed before and after treatment could be compared to assess behavioral changes that may have occurred as a result of sublethal pesticide exposure.

There are some problems inherent with the use of radiotelemetry to assess risk. These include proper species determination (what species are at risk), transmitter design, sample size, determination of the cause of mortality, and economic costs.

Species at risk

It is difficult to determine the wildlife species with greatest risk of pesticide exposure (and therefore targeted for intensive monitoring) without first obtaining adequate laboratory or LD50 data or having conducted a screening field test. For example, no data were available on the sensitivity of insectivorous birds to 1080 secondary poisoning, so potential hazards to these species could easily have been missed if not for the extensive carcass searches. Information on wildlife food habits and life histories is also important when determining the species to target for exposure and mortality monitoring during field studies. Three decisions made prior to initiating the studies discussed in this chapter illustrate this importance:

1) In the zinc phosphide orchard study, pheasants were known to be sensitive to zinc phosphide, were assumed to be at great risk, and were intensively monitored; however, few birds actually used the orchard habitat, preferring the weedy field borders instead.

2) California quail in the 1080 study were thought to be a good indicator species for monitoring seed-eating bird mortality; but during the late spring and early summer, quail did not consume seeds or grain baits and were therefore not exposed to the 1080 treatment.

3) In the strychnine pocket gopher study in western forests, chipmunks were identified prior to the study as a representative small mammal at risk from exposure to underground strychnine baits. However, they did not appear to forage extensively in the clearcuts; a later study (Anthony et al. 1984) showed greater risk to golden-mantled ground squirrels *(Spermophilus lateralis)*, a species not monitored in this study.

Transmitter reliability

A second problem with use of radiotelemetry is that transmitters are often unreliable for monitoring animals over periods long enough to assess hazards because of inadequate attachment techniques, size, and short signal longevity. Attachment of transmitters is difficult, particularly when applying them to birds. In these studies, we used harnesses, transmitters glued onto rubber patches that were then glued onto clipped feathers, tail clips, and other methods (Hegdal and Gatz 1976; Hegdal et al. 1986). Some methods caused abrasion of the skin; others allowed the transmitters to come off as feathers were molted or glue disintegrated.

Transmitter size is often a problem for small bird and mammal species, forcing use of transmitters as small as 1 to 2 g. It is common practice to limit transmitter weight to no more than 5% of a mammal's or 2 to 3% of a bird's weight (A. L. Kolz, personal communication). Even with small transmitters we have observed high mortality levels in pocket gophers and small birds, probably caused by transmitter weight. The process of instrumenting an animal could potentially make that animal more susceptible to pesticide effects or potentially cause behavioral responses that could attract predators. Therefore, transmitter-induced mortality makes it difficult to interpret risk data and forces the use of control areas when working with radio-equipped animals.

Longevity of the transmitter is a concomitant problem; because of size limitations for small bird and mammal transmitters, battery longevity may be only 2 or 3 weeks. Small bird transmitter batteries may fail after 1 week or 10 d. Few pesticide field studies can adequately assess risk within this short timeframe.

Sample size

Field risk assessment studies inherently suffer from sample size problems for several reasons: 1) initial populations of the animals targeted for monitoring may not be large, particularly for species with large territorial home ranges such as mammalian predators and raptors—in the California 1080 study, we were able to trap only 6 coyotes on the 90,000 ha study area; 2) trapping is frequently difficult—again, especially with predators and raptors—and requires experienced personnel and much time; 3) predator losses and migration can be significant for certain species and during certain times of the year; in the studies discussed, predator and migration losses often reduced the sample size monitored through treatment baiting, particularly for migratory birds, by at least 50%; and 4) when baiting occurs only in portions of the study area (such as in the zinc phosphide orchard study or strychnine pocket gopher studies), many of the radio-equipped animals may not use treated areas, further reducing sample size of exposed animals. For these reasons, it is difficult to obtain statistically acceptable sample sizes of transmittered animals.

Residue analysis

Determination of whether the pesticide contributed to mortality is not always easy, particularly when transmitters exert an effect on animals; therefore, a validated analytical method for animal tissue residues is necessary. In the 1080 study, the lack of adequate analytical methods for 1080 metabolites prevented us from stating conclusively that 1080 caused mortality to mammalian predators.

Cost and staff

A final problem with radiotelemetry studies is the high cost of equipment and staff. Transmitters and receivers are expensive and radiotelemetry studies are labor intensive. The studies described in this chapter would cost a minimum of $250,000 to $500,000 to conduct now. If multiple study sites were required, the cost would be multiplied accordingly.

Despite the many advantages of radiotelemetry, the problems listed above demonstrate the importance of using multiple means for assessing risk. Carcass searches, radiotelemetry, or other assessment techniques when used alone have been generally inadequate for developing quantitative risk assessments (RESOLVE 1994). The preceding studies show that when several techniques are used in combination, the risk assessment is greatly improved.

Acknowledgment- Four of these studies were funded by USEPA under Interagency Agreement EPA–IAG–4–449 between the USEPA and USFWS. Edward W. Schafer, Jr., and Craig A. Ramey reviewed the chapter.

References

Anthony RM, Lindsey GD, Evans J. 1984. Hazards to golden-mantled ground squirrels and associated secondary hazard potential from strychinine for forest pocket gophers. *Proc Vertebr Pest Conf* 11:25–31.

Atzert SP. Unpublished review of sodium monofluoroacetate—its properties, toxicology, and use in predator and rodent control. U.S. Department of the Interior, U.S. Fish and Wildlife Service.

Barnes Jr VG. 1974. Response of pocket gopher populations to silvicultural practices in central Oregon. Proc. Symp. on Wildlife and Forest Management in the Pacific Northwest, Corvallis, OR. p 167–175.

Barnes Jr VG, Anthony RM, Fagerstone KA, Evans J. 1985. Hazards to grizzly bears of strychnine baiting for pocket gopher control. *Wildl Soc Bull* 13:552–558.

Crouch GL, Frank LR. 1979. Poisoning and trapping pocket gophers to protect conifers in northeastern Oregon. U.S. Forest Service Research Paper PNW–261.

Fagerstone KA, Barnes Jr VG, Anthony RM, Evans J. 1980. Hazards to small mammals associated with underground strychnine baiting for pocket gophers. *Proc Vertebr Pest Conf* 9:105–109.

Fite EC, Turner LW, Cook NJ, Stunkard C. 1988. Guidance document of conducting terrestrial field studies. USEPA, Office of Pesticide Programs. Washington, DC. EPA 540/09–88–109.

Hansen RM, Ward AL. 1966. Some relations of pocket gophers to rangelands on Grand Mesa, Colorado. Colo. Agric. Exp. Sta. Tech. Bull. 88. 22 p.

Hegdal PL. Unpublished report to USEPA under Interagency Agreement EPA–IAG–D4–0449

Hegdal PL, Colvin BA. 1986. Radiotelemetry. In: Cooperrider AY, Boyd RJ, Stuart HR, editors. Inventory and monitoring of wildlife habitat. U.S. Dep. Interior, Bur. of Land Manage. Service Center, Denver.

Hegdal PL, Fagerstone KA, Gatz TA, Glahn JF, Matschke GH. 1986. Hazards to wildlife associated with 1080 baiting for California ground squirrels. *Wildl Soc Bull* 14:11–21.

Hegdal PL, Gatz TA. 1976. Hazards to wildlife associated with underground strychnine baiting for pocket gophers. *Proc Vertebrate Pest Conf* 7:258–266.

Hegdal PL, Gatz TA. 1978. Technology of radio tracking for various birds and mammals. Proc. PECORA IV Symp.: Application of Remote Sensing Data to Wildlife Management. Natl. Wildl. Fed. Sci. Tech. Ser. 3. p 204–206.

Johnson GD, Fagerstone KA. 1994. Primary and secondary hazards of zinc phosphide to nontarget wildlife—a review of the literature. DWRC Res. Rep. 11–55–005. Washington DC.: U.S. Dep. Agric., Denver Wildlife Research Center. 28 p.

LaVoie KL. Unpublished literature review and bibliography of strychnine. Denver CO: Denver Wildlife Research Center

Luce DG, Case RM. 1981. Damage to alfalfa fields by plains pocket gophers. *J Wildl Manage* 45:258–260.

Northwest Forest Pocket Gopher Committee. 1976. Survey of pocket gopher damage to conifers in the Pacific Northwest, 1975. Oregon–Washington Silvicultural Council, Western Forestry and Conservation Association, Portland, OR. 7 p.

Reeves HM, Geis AD, Kniffin FC. 1968. Mourning dove capture and banding. U.S. Fish Wildl. Serv. Special Sci. Report-Wildl. 117.

RESOLVE. 1994. Assessing pesticide impacts on birds: final report of the avian effects dialogue group, 1988–1993. Washington DC. 156 p.

Rudd RL, Genelly RE. 1956. Pesticides: their use and toxicity in relation to wildlife. Calif. Fish Game Bull. No. 7.

Tucker RK, Crabtree DG. 1970. Handbook of toxicity of pesticides to wildlife. U.S. Fish Wildl. Serv. Resour. Publ. 84.

Wood JE. 1965. Response of rodent populations to controls. *J Wildl Manage* 29:425–438.

Chapter 7

Radiotelemetry and GIS computer modeling as tools for analysis of exposure to organophosphate pesticides in red-tailed hawks

D. Michael Fry, Barry W. Wilson, Nancy D. Ottum, Julie T. Yamamoto, Robert W. Stein, James N. Seiber, Michael M. McChesney, and Elizabeth Richardson

Migratory and resident red-tailed hawks *(Buteo jamaicensis)* overwinter in the Central Valley of California, foraging and roosting in almond orchards. Application of organophosphate (OP) pesticides to orchards results in exposure to hawks; historical records document hawk mortality and birds exhibiting symptoms of OP poisoning. In this study, radiotelemetry, pesticide use reports, and geographical information system (GIS) mapping were used to analyze hawk home ranges. Footwash, feather residues, and plasma cholinesterases (ChEs) were studied to monitor exposure. Persistence of pesticide residues in orchards was determined from footwash data that were correlated with pesticide application date. Probable routes of OP exposure included dermal absorption through feet and ingestion of residues from prey and preening. Linear multiple regressions indicated ChE depressions of captured hawks were best correlated with footwash residues of parathion and less correlated with diazinon, methidathion, and chlorpyrifos. This study was sponsored by the Almond Board of California.

The Central Valley of California is a region of intense, mixed agricultural use including orchards, pastures, dairies, and row crops where resident and migratory hawks of several species overwinter. Red-tailed hawks are common, year-round residents. During winter months, the bird population increases with the arrival of migratory birds. Winter densities of red-tailed hawks were observed to be as high as 7/km² in some orchards. Since 1960, breeding populations have remained stable with yearly fluctuations, but winter populations have grown steadily (O'Connor unpublished data).

During winter, combined oil and organophosphate (OP) sprays are used to control San Jose scale *(Quadraspidiotus perniciosus)* and peach twig borer *(Anarsia lineatella)* in dormant almond and stone fruit orchards. Spraying dormant orchards reduces the total yearly application of OPs by as much as 40%. The major OPs used are parathion, diazinon, methidathion (Supracide®), and chlorpyrifos (Lorsban® or Dursban®). Before parathion sales were canceled in 1992, the application amounts were in the following order: parathion > diazinon > methidathion > chlorpyrifos. As the acres planted with almonds and stone fruits increased, raptors exhibiting signs of OP poisoning during the dormant spray season also increased. Wildlife incidence reports to California Department of Fish and Game (1962 to 1990) came from concerned individuals as well as Bidwell Nature Center, Feather River Wildlife Care, and Stanislaus Wildlife Care Center; these reports documented losses of birds and other wildlife around sprayed orchards.

Between 1987 and 1990, 16 red-tailed hawks were brought to wildlife care centers with symptoms indicative of OP poisoning, including depression, ataxia, tremors, lacrimation, and drooling. After administration of atropine, all 16 hawks recovered and were released. In addition, 18 dead hawks were reported in or adjacent to orchards during or directly following dormant-spray application (E. Littrell, personal communication).

Hooper et al. (1989) live trapped approximately 150 red-tailed hawks in almond orchard areas during 1986 and 1987 and confirmed through blood plasma ChE analysis that this species was being exposed to OPs during the dormant spray season. Exposures were sufficient for plasma ChEs to decrease in activity and to be reactivated with oximes specific for OP inhibition. In 1990, our state-mandated study was initiated to further assess the risks of OP exposure to red-tailed hawks.

Methods

The pesticides studied were ethyl parathion [O,O-diethyl O-(p-nitrophenyl) phosphorothioate], diazinon [O,O-diethyl O-(2- isopropyl-6-methyl-4-pyrimidinyl) phosphorothioate], methidathion [O,O-dimethyl phosphorodithioate S-ester with 4-(mercaptomethyl)-2-methoxy-1,3,4-thiadiazolin-5-one], and chlorpyrifos [O,O-diethyl O-(3,5,6-trichloro-2 pyridyl) phosphorothioate].

Animal care protocols were approved by the University of California-Davis campus veterinarian and the campus Animal Care Committee. Collection permits were obtained through the U.S. Fish and Wildlife Service (USFWS) and California Department of Food and Agriculture (CDFA).

Field studies

Site selection and description

This study was conducted during the winters of 1990 to 1991 and 1991 to 1992. An area of 127 km^2 southeast of Modesto, California, was selected as a study site based on its high proportion of almond orchards (36% of total area) and a history of OP-exposed red-tailed hawks being admitted to the nearby Stanislaus Wildlife Care Center in Ceres, California. The site was expanded to 147 km^2 by the end of the 1991 field season to allow for trapping of additional hawks and to account for movement of radioed hawks outside the edges of the original study area.

Land use maps of the field site were obtained from the California Department of Water Resources as hard copy overlays to U.S. Geological Survey (USGS) 7½-min quadrangle maps (scale 1:24,000). The maps were checked for accuracy and updated in both field seasons to take into account changing land use patterns such as removal or planting of orchards and construction of new houses.

CAMRIS® geographic information system

The Department of Water Resources land use maps were digitized using the CAMRIS geographic information system (Ecological Consulting, Inc., Portland, Oregon). Roads and canals were digitized from USGS 1:24,000 topographic maps. After digitizing was complete, 42 different land use types were identified. These land use categories were combined and simplified into 8 categories for most analyses: almond orchards, walnut orchards, peach orchards, other orchards (cherries, kiwis, apricots, apples), open fields (native pasture, alfalfa, fallow fields), other crops (corn, vegetables, vineyards), farmsteads (houses, commercial and urban areas, dairies, poultry facilities), and open water).

Sets of 12 overlapping, selected area land-use maps (1:15,000) were produced for ground truth editing and field observation entries. Corrections and additions to the land-use overlays were digitized directly from the ground-truthed 1:15,000 maps.

Trapping

Experienced hawk trappers were hired to live trap all possible red-tailed hawks *(Buteo jamaicensis)* and red-shouldered hawks *(B. lineatus)* within the study site. The primary methods used were bal-chatri traps (noose-covered, small, wire cages holding live prey), harnessed pigeon, and dhogaza mist nets (Berger and Mueller 1953). The study area was searched on a daily basis from 15 December 1990 to 15 February 1991, and from 22 November 1991 to 15 February 1992. Attempts were made to capture all hawks observed. Thirty-seven red-tailed hawks and 4 red-shouldered hawks were trapped and banded with USFWS lock-on bands. Several hawks of other species were inadvertently captured, including Cooper's hawks *(Accipiter cooperii)* and a ferruginous hawk *(B. regalis)*. These birds were banded, their feet were washed with ethanol to collect pesticide residues, 5 to 20 feathers were clipped per bird, and the birds were immediately released.

Sample collections

Trapped hawks were hooded and transported by car to a field laboratory. The feet of each bird were rinsed thoroughly with 95% ethanol; the rinsings were collected into solvent-rinsed bottles and stored on dry ice for pesticide residue analysis. All birds were weighed and wingspread measured from tip to tip. A blood sample of up to 1.0 ml was taken from the brachial or metatarsal veins, centrifuged, and separated into aliquots for clinical chemistry and ChE measurements. Blood smears and plasma (stored on wet ice) were sent to California Veterinary Diagnostics, Inc., West Sacramento, California, for hematological and serum chemistry analyses. Within 24 h of plasma delivery, enzyme activities and chemical analyses were determined using standard methods with a Hitachi 736 Autoanalyzer. Plasma samples were frozen on dry ice for ChE determinations. After measurements and blood samples were collected, radio transmitters were attached to birds, and they were returned to the transport crate and held 2 h to overnight for collection of a fecal sample. The hawks were fed and released at the capture site.

Radio transmitters

Two-stage radio transmitters were obtained from Communication Specialists, Irvine, California, and from Holohil Systems Limited, Woodlawn, Ontario, Canada. Frequencies used were within the 148 to 150 MHz range. In the first season, 11 red-tailed hawks were each fitted with a transmitter weighing 24 g (Communication Specialists) with a designed battery life expectancy of 21 months. Seven red-tailed hawks and 2 red-shouldered hawks were each fitted with a transmitter weighing approximately 17 g with a 4-month battery life (Communication Specialists). During the second season, 18 red-tailed hawks and 2 red-shouldered hawks were each fitted with a 14 g, 23-month transmitter from Holohil. One red-tailed hawk was caught too late in the season to fit it with a transmitter.

The transmitters were attached to the hawks with 6-mm wide, braided teflon ribbon in a backpack configuration. The ribbons were crimped together with a brass ferrule at the bird's breast. The 34 cm-long antennas extended just beyond the tips of the retrices.

Radio transmission was monitored with hand-held, 3-element yagi antennas (Communication Specialists) and ICOM Model IC-H16 synthesized frequency transceivers (Communication Specialists). The 16 channel, programmable transceivers were also used as 2-way radios at 154.57 MHz under a Federal Communications Commission license.

The detection range of transmitters varied under field conditions from 0.5 km to as much as 15 km, depending upon whether the bird was in a dense orchard or soaring at 100 to 300 m above the ground. Some birds twisted the braided antenna wire into knots with their beaks, causing the signal to become greatly attenuated. During the second season, shrink-wrap tubing placed over a flexible antenna greatly reduced kinking and knotting.

Field radio tracking

Between December and 15 February, field teams of 2 to 6 people searched the study area several times per week to locate and observe radio-tagged hawks. Birds were located visually whenever possible, but location by triangulation from 100 to 200 m was occasionally necessary in orchards. Trees used as night roosts were identified by triangulation or direct observation of birds flying to them. The location of each observation was marked on overlay maps and recorded on data sheets with date, time, location description, and behavioral notes.

Geographic information system data analysis

Hawk observations were digitized directly from overlay maps into CAMRIS, and data sheet information was entered into DBase III© (Ashton-Tate). The database files and GIS program were linked for data analysis by using unique identification numbers for each observation point. The number of observation points varied from as few as 1 to more than 80 per bird. Five birds immediately left the study area after release and were not detected again.

The areas used by hawks were evaluated with CAMRIS using several comparative techniques. Observation datapoints were overlaid with land uses, and the number of observations in each land use (habitat type) were summed. Home ranges were computed using a density surface contour program in CAMRIS (Ford and Krumme 1979) by overlaying a grid (400-m spacing for red-tailed hawks, 200-m spacing for red-shouldered hawks) on the observation points which CAMRIS then used to calculate density contours around the clusters of datapoints. A contour line surrounding 100% of the observation points represented the entire home range, while the density contour surrounding the most tightly clustered 50% of observation points was considered the core area of the home range. The habitat use by the hawk was computed by measuring the overlap of the home range polygons with land use polygons.

Determinations of home ranges using the minimum convex polygon method (Mohr 1947) and a harmonic mean method adapted from Dixon and Chapman (1980) were rejected because they did not provide home range estimates that fit the observed land use of hawks as well as did the density surface contour method (Ford and Krumme 1979).

Some hawks did not remain in a fixed area, but wandered throughout large areas of the study site during the winter. No reliable home range could be computed for these birds, and they were excluded from an analysis of pesticide exposure risk. Birds with fewer than 15 locational datapoints were also excluded from analysis because too few datapoints were collected to adequately describe the home range (Ford and Meyers 1981).

Pesticide use reports

The State of California requires growers to submit intent-to-spray forms to the County Agricultural Commissioner within 30 d of pesticide application and to file a use report after application. Each use report contains date and time of spraying, application rate, name of pesticide, and a sketch map of the area. Copies of permits and use reports were obtained from the Stanislaus County Agricultural Commissioner for all applications of parathion, diazinon, chlorpyrifos, and methidathion in the winter of 1990 to 1991 and 1991 to 1992. Sketch maps were compared to the digitized land use overlays, and the acreage sprayed was entered onto overlays in CAMRIS. The acreage of each plot was verified during entry. Each spray plot polygon in CAMRIS was given an identification number corresponding to an entry in DBase III containing the pesticide, date of application, permittee, and acreage sprayed.

Evaluation of exposure risk

Home-range polygons were overlaid on pesticide spray maps to determine the percent of home range sprayed, the types of pesticides sprayed, and the number of days prior to hawk capture that spray applications occurred. The percent of home range sprayed provided a relative assessment of probable exposure.

Within each bird's home range, data from residue analysis of footwashes and feathers were correlated with number of days since pesticide application to calculate a measure of pesticide persistence within the orchards.

Residue analyses and enzyme assays

Organophosphate parent compounds (chlorpyrifos, diazinon, methidathion, and parathion) and their oxons were analyzed from footwashes and feather samples as in Seiber et al. (1989), Glotfelty et al. (1990), and Hooper et al. (1989). ChE activity in plasma was determined by the colorimetric method of Ellman et al. (1961) and modified for use with an automated microplate reader. The nonspecific cholinesterase (butrylcholinesterase [BChE]) inhibitor, iso-OMPA, was used to distinguish between BChE and acetylcholinesterase (AChE).

Plasma enzyme and organophosphorus correlations

Stepwise linear regression was used to examine correlations of footwash residues with blood cholinesterase activities for red-tailed hawks using Statistical Analysis System (SAS) (Proc REG, option = stepwise). Linear regression equations were calculated for Total ChE, AChE, and BChE activities with individual OPs considered simultaneously to combine their relative contributions to blood ChE depression.

Results

Land use patterns and pesticide applications

The initial field site selected was 127 km^2 of flat, valley bottomland consisting of 36% almond orchards (4,600 ha). In the second field season, the site was enlarged to 147 km^2 with 39% almond orchards (approximately 5,700 ha) (Table 7-1). Enlarging the site did not markedly affect the land use patterns. The acreage added was slightly rolling hills with larger blocks of orchards planted to single crops than in the original site. The additional orchards were mostly sprinkler-irrigated, while flood irrigation was used in much of the original plot.

Pesticide use varied between the years, especially because of the cancellation of parathion sales after 31 December 1991. The amount of parathion applied in the winter of 1992 was much reduced compared to 1991, and there was a corresponding increase in the use of methidathion, diazinon, and chlorpyrifos (Table 7-2). The size of the blocks sprayed differed between pesticides; the average plot sprayed with methidathion and chlorpyrifos was 8 ha, while the average block size for parathion was 32 ha. After parathion sales were canceled, several large growers shifted to methidathion, resulting in an increase in block size for methidathion in the second season.

Radiotelemetry, land use, and hunting behavior

Thirty-six red-tailed hawks were fitted with radio transmitters (Table 7-3), of which 25 remained in the study area and were observed 15 times or more within 2 weeks following their capture and release. Four of the adult red-tailed hawks trapped in the first season remained in the study area throughout both years of the study. Two wintering birds returned to the study site in the second year. Additional birds returned with nonfunctional radio transmitters; attempts to retrap these birds were not successful.

The 2 red-shouldered hawks fitted with short-lived (4 to 5 month) radio transmitters in 1990 to 1991 remained in the area during the breeding season. Banded red-shouldered hawks with attached, non-functioning transmitters were occasionally observed in the appropriate locales throughout the study, suggesting hawks trapped in the first year remained throughout both years.

Habitat utilization varied greatly among individual hawks. Several red-tailed hawks spent the day hunting in open pastures and perching on telephone poles or fences, moving to large trees or willow thickets at dusk to roost. Other red-tailed hawks stayed in the almond orchards, perching in trees and on telephone poles within the orchards. Most red-tailed hawks spent much of the day perched on telephone poles near the edges of orchards, probably hunting in the grasslands and pastures as well as in the orchards. The red-tailed hawks hunted by first scanning a large area from prominent perches and then employing long glide-dives to attack prey. Qualitative observations indicated that most attacks on natural prey occurred 100 m or less from the perch. Red-tailed hawks attacked small mice in bal-chatri traps set 200 m from their perches.

Table 7-1   Major land use types of red-tailed hawk telemetry study area, winter 1990 to 1991 and 1991 to 1992

Land use	Size (ha)		Percent	
	Year 1	Year 2	Year 1	Year 2
Almonds	4,600	5,700	36.00	39.00
Walnuts	770	690	6.10	4.70
Peaches	800	780	6.30	5.30
Other orchards	42	82	0.30	0.60
Other crops	2,740	3,140	22.00	21.00
Open fields	3,040	3,270	24.00	22.00
Farmsteads	680	1,050	5.40	7.10
Water	22	24	0.20	0.20
Total	12,694 ha	14,736 ha	100.3%	99.9%

Table 7-2   Pesticides applied on orchards for both field seasons of red-tailed hawk telemetry study (ha)

Pesticide	1990–91	1991–92
Parathion	995	274
Diazinon	229	656
Chlorpyrifos	347	601
Methidathion	970	1,105
Total	2,541 ha	2,636 ha
% Study area	20%	18%

Most red-shouldered hawks preferred to remain within almond and walnut orchards, although they were occasionally observed on the edge of the orchards. Some used eucalyptus trees and riparian (willow and oak) trees adjacent to pastures in addition to trees in the orchards.

Table 7-3   Numbers of hawks trapped and radio tagged for telemetry field study near Modesto, California, during the winters of 1990 to 1991 and 1991 to 1992

1990 to 1991	
18 red-tailed hawk	12 adults and 6 immatures
2 red-shouldered hawk	2 adults
1 retrapped red-tailed hawk	1 adult
1991 to 1992	
18 red-tailed hawk	14 adults and 4 immatures
2 red-shouldered hawk	2 adults
3 retrapped red-tailed hawk	3 immatures

Home ranges and hawk interactions

The home ranges of red-tailed hawks varied from 79 to 329 ha, averaging 181 ha. The territories of red-shouldered hawks were much smaller, averaging only 69 ha. Home-range size of resident red-tailed hawks (birds remaining within a study site throughout the breeding season that followed the spray season) were not significantly different from those of wintering birds (birds not remaining for the breeding season) with stable home ranges. Although the home range locations of resident red-tailed hawks overlapped very little with those of wintering birds, the areas used by many wintering birds overlapped each other. This implies that an active exclusion of other red-tailed hawks was practiced by resident birds, even though very few aggressive interactions were noticed during the study.

Two pairs of red-tailed hawks had both members fitted with transmitters. In both cases, males and females appeared to use the same territories. Paired birds often roosted together in the same or adjacent trees, and they perched by themselves during the day. One pair of wintering red-tailed hawks had overlapping territories in the first season; they left the study area in the spring, and only 1 bird returned to the same area in the second year. Only about half of the wintering red-tailed hawks established stable home ranges; others wandered through the study area. Two wandering red-tailed hawks were resident birds that remained in the study area throughout the spring and summer.

Home ranges and spray applications

Figures 7-1 through 7-3 map 3 red-tailed hawk home ranges and are overlaid with polygons that represent sprayed orchards. Numbers within polygons represent the number of days between when the orchard was sprayed and when the hawk was captured. Footwash residues recovered from the birds at capture (total µg per bird) are given at the bottom of the figure legend. Red-tailed hawk band #302 had no residues on its feet when first captured on 20 December 1990 (Figure 7-1). However, it had footwash residues of 4 OPs when recaptured on 23 January 1991 (#302 was the only bird captured twice in the 1990 to 1991 field season). Several orchards within the home range of red-tailed hawk #302 were sprayed prior to recapture. The orchard in which it was retrapped had been

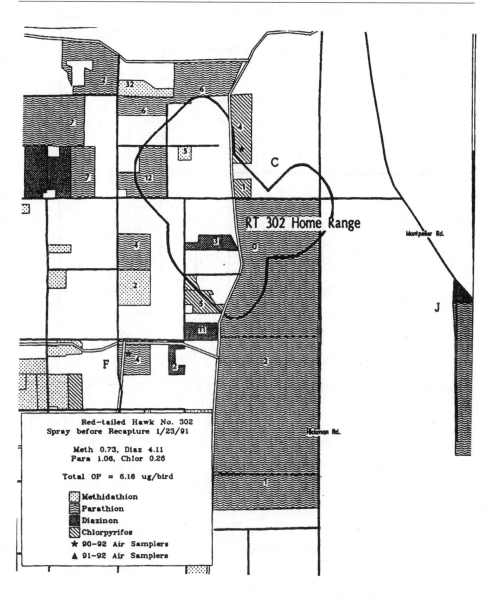

Figure 7-1   Home range of red-tailed hawk #402 and areas of spraying prior to bird's recaptue. Note: Numbes within polygons represent the number of days between when the orchard was sparyed and the hawk was captured. Footwash resideues recovered from the birds at capture (total µg per bird) are given in the figure legend.

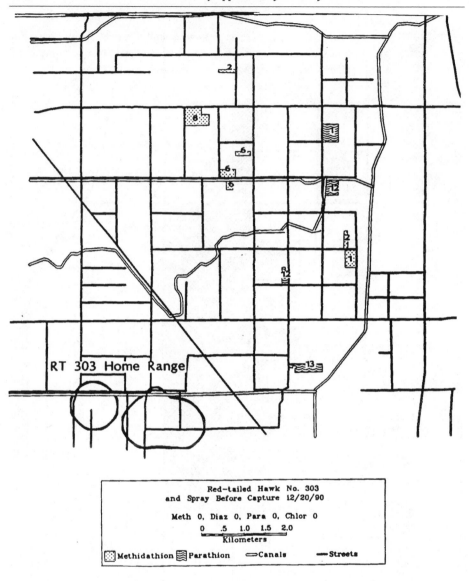

Figure 7-2   Home range of red-tailed hawk #303 and areas of spraying prior to bird's initial capture. Note: Numbers within polygons represent the # of d between when the orchard was sprayed and the hawk was captured. Footwash residues recovered from the birds at capture (total μg per bird) are given in the figure legend.

sprayed with diazinon 3 d before. Other orchards were sprayed with chlorpyrifos 1, 4, and 5 d before; methidathion 5 d before; and parathion the day the hawk was recaptured. Since residues of all 4 compounds were recovered from its feet, it is apparent that the hawk moved extensively throughout its home range.

Figure 7-2 presents similar data for red-tailed hawk #303, a bird that had no measurable OP footwash residues when captured on 20 December 1990. It was trapped relatively early in the spray season when only a few orchards had been sprayed, and none had been sprayed within its home range. Similar data are presented in Figure 7-3 for red-tailed hawk #740; only parathion was recovered from its footwash, and parathion had been applied within its home range 10 d prior to capture. Diazinon had also been applied 10 d before, and methidathion was applied 7 d before capture. Either the hawk did not use the latter orchards after they were sprayed, or the pesticides were rapidly lost from the trees or the feet of this hawk.

Pesticide exposure of hawks

During the dormant season, pesticide applications in almond, peach, and walnut orchards were limited to November through January. In both years of the study, only 1 of 15 hawks captured during the months of November and December had more than 1 μg total OP residues on its feet (Figure 7-4). In January and February, 21 of 26 birds had total residues greater than 1 μg. A total of 37 hawks were captured with 4 birds being recaptured at least 30 d later and sampled by footwash both times. Recapture data is included in the results and assumed to be independent of the previous footwash. More than 1 compound was detected on 20 of 22 hawks sampled, indicating that the birds moved through several different orchards after spraying. Nine red-tailed hawks had detectable OP residues on their feathers. Two of the 4 red-shouldered hawks had OP residues greater than 1 μg on their feet.

Mortality

Four radio-tagged red-tailed hawks died by various causes: 1 was shot on a poultry ranch, 1 was electrocuted, 1 was hit by a car adjacent to an orchard sprayed with parathion 19 d previously, and 1 was found decomposed in a peach orchard 13 d after application of guthion on June 24. The latter bird was observed alive on 24 June flying above the orchard. A radio signal, but no visual observation, came from the orchard on 1 July. By 8 July, only skeletal remains were found.

Cholinesterase levels and correlations with residues

Serum cholinesterase depression was observed in many exposed hawks. Of the 37 birds sampled, 13 had AChE levels and 3 had BChE levels depressed at least 2 standard deviations below the means described in our previous study of birds captured in November and December 1986 and 1987 (Hooper et al. 1989), levels we have operationally defined as indicating significant exposures had occurred. BChE was responsible for a large percentage of total ChE activity in red-tailed hawks, in agreement with previous studies from

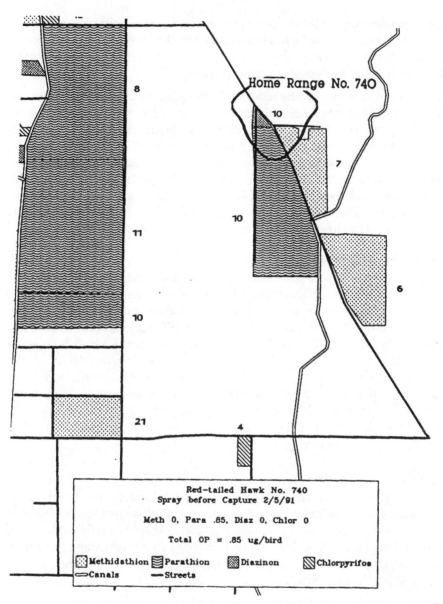

Figure 7-3   *Home range of red-tailed hawk #740 and areas of spraying prior to bird's initial capture. Note: Numbers within polygons represent the number of days between when the orchard was sprayed and the hawk was captured. Footwash residues recovered from the birds at capture (total μg per bird) are given in the figure legend.*

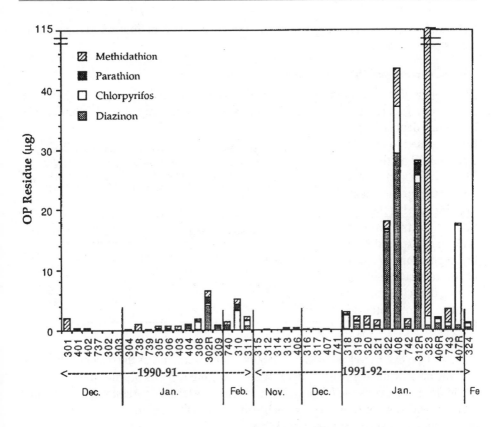

Figure 7-4   *OP residues in footwashes of red-tailed hawks. Birds are listed by band number in order of capture for each field season (1990 to 1991 and 1991 to 1992). Recaptured birds are indicated by "R" after band number.*

our laboratory (Wilson et al. 1992). No bird from either field season had 2-PAM reactivable ChE activities.

Regression analyses are presented for birds with more than 1 g total OP residue in their footwash (Table 7-4). These equations describe the change in cholinesterase activity from a mixture of OP residues. The equations are of the form

$$ChE = Intercept + X(PA) + Y(DZ) + Z(ME) + W(CP) \qquad \text{Equation 7-1.}$$

ChE is cholinesterase activity. Intercept is the estimate of ChE activity in the absence of OP residues. PA, DZ, ME, and CP are µg of footwash residues of parathion, diazinon, methidathion, and chlorpyrifos, respectively. W, X, Y, and Z are factors representing the regression line and, therefore, cholinesterase depression (–) or induction (+) above 0 residues. Insofar as depressed blood ChEs represent an estimate of potential toxicity of an

OP pesticide, the larger a negative factor, the greater its contribution to the risk the sprays pose to hawks.

Plasma enzyme and OP correlations

Correlations of footwash residues with blood cholinesterase activities (Table 7-4) indicate that 3 pesticides (parathion, methidathion, and diazinon) contributed to blood cholinesterase depressions for birds exposed to at least 1 g of total OP on their feet. Chlorpyrifos residues did not contribute sufficiently to be included in the equations when processed by SAS. In unexposed birds, the regression equations are quite accurate in estimating the activity of each of the cholinesterases. Hooper et al. (1989) gave the unexposed ChE level as

Table 7-4   *Stepwise linear regression correlating footwash residues of parathion (PA), diazinon (DZ), and methidation (ME) with blood cholinesterase activities for red-tailed hawks (RTHA), which had more than 1 µg of OP residues in their footwash*

Regression equation	R squared	N
ChE ɜ 0.71 & 0.18PA & 0.0094DZ & 0.004ME	0.58	19
BChE ɜ 0.56 & 0.15PA & 0.0069DZ & 0.0028ME	0.50	19
AChE ɜ No equation		

0.790 and the BChE level as 0.488 µmoles/ml/min. Parathion was the major contributor to ChE and BChE depression here, with diazinon responsible for only 5%, and methidathion responsible for only 2% of the cholinesterase depressions. SAS did not derive an equation for AChE depression because no variable met the $P < 0.15$ significance level required for entry into the model.

Discussion and conclusions

Limitations of the techniques

The 3 major limitations to this study were the following: 1) difficulty in trapping red-tailed hawks; 2) reliance on reporting of pesticide applications by growers; and 3) physical limitations of power, weight, and battery life of radio transmitters. The field study showed that wintering red-tailed hawks tended to remain in specific home ranges that varied greatly in their composition. Land use by birds varied from exclusive use of orchards to no use of orchards. In almost every case, hawks were exposed to pesticides whenever an orchard within their home range was sprayed. The presence of residues from more than 1 pesticide in the footwashes and feathers indicated that hawks did not avoid sprayed orchards. This is confirmed by our direct field observations of hawks flushed from orchards by spray rigs and their immediate return after the rig had passed. There was 1 record of a red-tailed hawk consuming avian prey (killdeer, *Charadrius vociferus*) in a recently sprayed orchard. Although the study did not directly establish whether ingestion of exposed prey posed an increased risk to hawks, secondary toxicity from ingestion of exposed prey has been demonstrated in other studies (Hunt et al. 1991).

The inability to retrap red-tailed hawks required us to obtain the pesticide residue samples at the time of initial capture and to perform a retrospective study of pesticide exposure based on the random capture dates. While this provided for a completely un-biased sample of hawks during exposure, it presented significant problems in both field logistics and data analysis. Only those birds remaining in the area of capture with a stable home range could be used for pesticide exposure risk analysis. Since the hawk capture and pesticide applications were not coordinated, a complete and timely report-ing of pesticide use was required to have accurate data. This posed some difficulty, as the researchers had only limited ability to independently verify the accuracy of pesticide use reports. We used our field observations of spray activities that were recorded incidentally to hawk observations, and we maintained a set of 12 pesticide air sample monitors throughout the study area as an accessory to this study. We believe the reporting by growers within the study site was highly accurate.

The third limitation may be a major difficulty in a large area of study with highly mobile birds. The inability to reliably detect birds at more than 1 to 2 km required extensive field monitoring with a great deal of travel time within the study site. When birds moved a few kilometers away from their previous known location, finding them was nearly a random process.

Land use and home range analysis

The patchy and rectilinear nature of habitats used by hawks in this study required com-parison of several analysis methods for determination of home range. The minimum convex polygon method (Mohr 1947) constructs a home range by connecting the most peripheral observation points into the convex polygon of smallest size possible. Two obvious drawbacks to this method are apparent. First, areas within the polygon that are not used by the hawk will be included as part of the home range. Second, for example, a pasture adjacent to an orchard that is hunted from an orchard perch would not be in-cluded unless the bird was actually observed flying into the field and landing. One vari-ant of the minimum convex polygon method can be the construction of a buffer zone of fixed width (e.g., 200 m) around the polygon to accommodate this hunting behavior, but the drawback of inclusion of dead areas within the polygon cannot be avoided.

The harmonic mean method of analysis (Dixon and Chapman 1980) uses a formula that assigns grid nodes a numerical value derived as the sum of reciprocals of the distance from every observation datapoint. Since each observation point is related to every other observation, dead areas within the home range will be included but can be minimized by weighing coefficients incorporated into the calculation formulas. The process of weigh-ing is largely empirical, however, and different combinations of weighing coefficients are used to fine tune the home ranges so points and clusters of points result in high numerical values. A contour surface of probabilities is generated using the numerical values, and the 95% or 99% probability contour can be defined as the home range.

The density contour method selected in this study requires the selection of a grid spacing that has a large effect on the resulting contours. A grid spacing of 400 m was used for red-tailed hawks based on the observations that prey were observed and pursued when placed at 200 m from a perched bird. This implies a 400-m-diameter circle of hunting activity, making a 400-m grid spacing in the analysis appropriate. A 200-m grid spacing was selected for red-shouldered hawks because they hunt primarily within orchards and do not usually have as large an area to observe.

For most red-tailed and red-shouldered hawks, both the harmonic mean and density surface contour methods produced similar home range results when large weighing coefficients were used in the harmonic mean calculations. The density surface contour method was selected because of its simplicity and lack of need for arbitrary weighing coefficients to produce realistic results.

This study demonstrates the power of a geographical information system to integrate diverse information and to provide temporal and spatial analysis of several datasets. The combination of land use data, hawk radiotelemetry observations, pesticide application information, and biochemical measures of toxicity enabled us to demonstrate and assess pesticide exposure. Exposure risk could be determined through the use of multiple regressions comparing dermal exposure (footwash residues) with serum cholinesterase depression. The pesticide use information correlated with exposure data provides a measure of exposure risk duration for hawks entering orchards after spraying, and significant residues could be detected on birds for up to 2 weeks in many instances.

In addition to the important exposure-risk parameters determined for each of the 4 pesticides, several unexpected facets of hawk behavior were revealed by CAMRIS analysis, including home-range size, territoriality of resident but not overwintering hawks, and the diversity of habitats used. Thus, the combination of radiotelemetry, GIS data analysis of home ranges, and pesticide-use data provided significant new information on pesticide exposure risk to birds using orchard habitats.

Acknowledgments - We would like to thank Peter Bloom, Ed Henckel, and Judy Henckel for trapping hawks, and Donna Burt, Stanislaus Wildlife Care Center, Ceres, California, for assisting in blood sampling and care of injured birds. Tom Fraser, Jeff Fraser, Mike Morgan, Molly Flemate, Fern Williams, and Patty Healey provided valuable help as volunteers. Denise Chakerian provided valuable technical editing assistance. This study was supported by the Almond Board of California.

References

Berger DD, Mueller HC. 1953. The bal chatri: a trap for the birds of prey. *Bird-Banding* 30:18–26.

Dixon KR, Chapman JA. 1980. Harmonic mean measure of animal activity areas. *Ecology* 61:1040–1044.

Ellman GL, Courtney KD, Andres V, Featherstone RM. 1961. A new and rapid colorimetric determination of acetylo-cholinesterase activity. *Biochem Pharm* 7:88–95.

Ford RG, Krumme DW. 1979. The analysis of space use patterns. *J Theor Biol* 76:125–155.

Ford RG, Meyers JP. 1981. An evaluation and comparison of technique for estimating home range and territory size. In: Scott JM, Ralph CJ, editors. Estimating the numbers of terrestrial birds, Vol. 6—studies in avian biology. Los Angeles CA: Cooper Ornithological Society. p 461–465.

Glotfelty DE, Schomburg CJ, McChesney MM, Sagebiel JC, Seiber JN. 1990. Studies of the distribution, drift and volatilization of diazinon resulting from spray application to a dormant peach orchard. *Chemosphere* 21:1303–1314.

Hooper MJ, Detrich PJ, Weisskopf CP, Wilson BW. 1989. Organophosphorus insecticide exposure in hawks inhabiting orchards during winter dormant-spraying. *Bull Environ Contam Toxicol* 42:651–659.

Hunt KA, Bird DM, Mineau P, Shutt L. 1991. Secondary poisoning hazard of fenthion to American kestrels. *Arch Environ Contam Toxicol* 21:84–94.

Mohr CO. 1947. Table of equivalent populations of North American small mammals. *Am Midl Nat* 37:223–249.

O'Conner R, University of Maine. Unpublished data.

Seiber JN, McChesney MM, Woodrow JE. 1989. Airborne residues resulting from use of methyl parathion, molinate and thiobencarb on rice in the Sacramento Valley, California. *Environ Toxicol Chem* 8:577–588.

Wilson BW, Hooper MJ, Hansen ME, Nieberg PS. 1992. Reactivation of organophosphate inhibited AChE with oximes. In: Chambers JE, Levi PE, editors. Organophosphates: chemistry, fate and effect. San Diego CA: Academic.

Chapter 8

Use of radiotelemetry to monitor survival of songbirds exposed to an insecticide on golf courses

Richard M. Poché, David L. Fischer, and Paul A. Toll

As part of a terrestrial field study to evaluate the potential for acute avian mortality resulting from applications of a new insecticide, survivorship of ground feeding birds was monitored using radiotelemetry at 8 golf courses in Columbus, Ohio. A total of 560 songbirds (497 American robins [*Turdus migratorius*], 46 blue jays [*Cyanocitta cristata*], and 17 brown thrashers [*Toxostoma rufum*]) were captured in mist nets 3 to 21 d prior to scheduled chemical applications and fitted with radio transmitters weighing about 1.5 g each. Transmitters were attached above the birds' synsacrums and held in place with strands of nylon elastic cord looped around each leg. Attachment of the transmitter required only 1 to 2 min and did not appear to negatively affect behavior of these species. Each golf course was divided into 2 plots, 1 of which was randomly selected to receive the insecticide treatment. Radio signals from radio-tagged birds were monitored from a minimum of 2 stations per plot. Records were maintained on direction of signals, intensity of signals, and movement detection. A moving (and therefore alive) bird produced a radio signal of fluctuating intensity. When the signal intensity and direction appeared constant, an effort was made to find the bird via radio tracking and determine its fate. Of 560 birds equipped with transmitters, 500 (89%) were detected alive and present on study plots just prior to application. Of these, 423 (85%) were known to survive until at least 7 d postapplication when the chemical no longer posed a risk. The survival index at insecticide-treated plots (89%) was higher than the index at control plots (80%), indicating the chemical application had little if any adverse effect on survival of the focal species.

Radiotelemetry has been demonstrated to be useful in many types of wildlife investigations. Numerous transmitter attachment designs have been developed and field tested (Raim 1978; Cochran 1980; Odom et al. 1982; Mech 1983; Rappole and Tipton 1991). The study described herein demonstrated the use of radiotelemetry to evaluate the survival of songbirds exposed to pesticide applications.

Purpose

This terrestrial field study was conducted to evaluate the acute hazard to birds posed by applications of a granular insecticide to turf grass on golf courses. Based on laboratory data, ingesting 50 to 100 pesticide granules—fewer than the number of granules applied to each square foot of the treatment area—would be potentially lethal to birds. Survival of radio-tagged birds potentially exposed to the application was monitored using radio-telemetry. Prior to the conduct of the field study, a preliminary study was conducted to 1) determine efficiency of transmitter attachment on passerines using 4 harness designs, 2) test transmitters from 3 manufacturers to select the appropriate type for robin-sized passerines, and 3) determine the approximate longevity of the reported 60-d batteries.

Background

The field study was conducted to evaluate the acute risk to birds posed by applications of an experimental granular insecticide. Results of laboratory studies indicate this insecticide is moderately toxic to quail and mallard ducks, but it is highly toxic to songbirds.

Birds may be exposed to granular pesticides by ingesting granules clinging to other food sources or by picking up the granules as a source of grit. Therefore, the species of birds at greatest risk for this formulation type are ground-foraging passerines such as the American robin (*Turdus migratorius*).

Methods

Preliminary study

A preliminary aviary study using American robins was conducted in St. Martin Parish, Louisiana, during December 1991 and January 1992 to determine the optimum transmitter harness design and transmitter type to be used on passerine birds in the proposed field study. Four transmitter harness designs and 3 transmitter types were tested on American robins. The transmitter manufacturers were each instructed as to the weight limitations of the transmitters and the designs of the 4 different harnesses to be tested. The manufacturers used were 1) Wildlife Materials, Inc., Carbondale, Illinois, 2) Holohil Systems, Ltd., Woodlawn, Ontario, and 3) ATS, Inc., Isanti, Minnesota.

The 3 transmitter types ranged in weight from 1.4 to 4.11 g, with Holohil transmitters being the lightest and ATS transmitters being the heaviest. All transmitters had an estimated 60-d battery life. The frequency of these transmitters was 150 to 151 MHZ.

Two harness designs tested utilized the leg-loop attachment technique described by Rappole and Tipton (1991). The first of these used a non-flexible loop made of nylon fishing line epoxyed into the base of the transmitter, and the second used a flexible loop made of 1.5-mm nylon elastic cord (loop size about 62 mm tip to tip) tied to fit the size of the bird. A breast-mount harness was the third design tested, which secured the transmitter to the robin's breast with a loop over the head and 2 elastic bands placed under the wings and over the back. The fourth harness design used string placed under the wing and over the breast to secure the transmitter, which was glued on the lower back above the tail.

The aviary used to house the birds was approximately $30.5 \times 30.5 \times 6.1$ m, constructed of plastic netting, cables, and oilfield pipe. The netting (19×32-mm mesh) was draped over the cables attached to the pipes.

We used mist nets to capture American robins in the early evening in abandoned sugar-cane fields. The robins were removed from the nets, placed into cloth bags, and then transferred to 0.6 × 0.6 × 0.3-m cardboard cartons for immediate transport to the aviary. The birds were fitted with 1 of 3 radio transmitter types attached by 1 of 4 harness types prior to being released into the aviary.

The captive birds were provided holly berries, blueberries, soft china berry seeds, boiled rice, boiled chicken eggs, a special soft-billed bird pellet (Reliable Protein Products, Rancho Mirage, California), and water ad libitum.

Field study

The field study was conducted between April and July 1992. This period corresponds to the breeding season; therefore, birds are more sedentary and the insecticide would normally be applied during this time of year to control turf pests.

A single application of the insecticide at the maximum label application rate was made to 8 golf courses in Franklin and Delaware Counties, Ohio. Each golf course was randomly divided into a treatment and control plot. Survival of marked birds was monitored using radiotelemetry.

Study site

Eight golf courses were selected in Columbus, Ohio, for the study. Each course was divided into randomly selected treatment areas and one nontreated control area, based on the distribution of songbird habitat, layout of the fairways, and size of the course. A buffer zone (minimum of 1 to 2 fairways) separated the two plots. The golf courses selected were Ohio State University Golf Course, Columbus Country Club, Minerva Lake Golf Course, Hickory Hills Golf Club, Oakhurst Country Club, Blackhawk Golf Course, Champions Golf Course, and Brookside Country Club.

Netting and banding

American robins, blue jays (*Cyanocitta cristata*), and brown thrashers (*Toxostoma rufum*) were the focal species selected for the radiotelemetry portion of this study. A target of 30 robins, 5 blue jays, and 5 brown thrashers tagged per plot was established as our marked population.

Thirty mist nets per plot were used to capture the birds. When one of the focal species was captured, the bird was removed from the net, banded with a U.S. Fish and Wildlife Service-numbered leg band, given a unique color identification using colored leg bands, fitted with a radio transmitter emitting a unique frequency, and released.

Transmitters and harness

Transmitters selected for the field study were manufactured by Holohil Systems, LTD, and Wildlife Materials, Inc. (see results). These transmitters weighed approximately 1.5 g, were in the frequency range of 148.0 to 152.0 MHZ, and had an estimated receiving range of 1 mile. The pulse rates were 45 to 60 pulses per min. The battery life was projected to be a minimum of 60 d.

The harness best suited for the birds was a modified leg-loop design (Rappole and Tipton 1991) utilizing a 1.5-mm nylon elastic cord inserted through 2-mm plastic tubes, embedded in epoxy at opposite ends of the transmitter. Holohil embedded the loop ends in the epoxy base of the transmitter, and Wildlife Materials, Inc. provided the cord, which was tied to each unit prior to going into the field. This method of attachment enabled field personnel to attach the transmitter to the bird within 1 to 2 min.

Tracking radio-tagged birds

Data were collected on movement of individual birds to determine survival. On days −7, −3, 0, 3, 7, and 10 postapplication, transmitter signals were monitored using TRS-2000 receivers (Wildlife Materials, Inc.) and a 3-element, folding, directional yagi antenna.

To track bird movements, a minimum of 2 tracking stations were established on each plot. Each station was established by placing tent pegs marked with compass coordinates in a 3-m-diameter circle around the tracker. This layout enabled the tracker to obtain a quick directional bearing once the signal had been detected. Directional bearings were recorded for plotting bird locations on plot maps.

Simultaneous bird locations were obtained at each plot by technicians communicating via radio. Two attempts were made during each tracking session to locate all radio-tagged birds released on each plot. A bird was classified as alive if its position changed or its transmitter signal fluctuated, indicating movement. When a signal was received only during the second location attempt, we assumed the bird had moved from a previously undetected location and was, therefore, alive. Signals that were not received during either attempt and signals that remained stationary during both location attempts were chased down by the technicians by driving in golf carts to the signals to determine the status of the birds. The areas around the golf courses were monitored from preselected sites in an attempt to locate radio signals not detected on the study plots. This monitoring was conducted from a vehicle equipped with a roof-mounted, omni-directional antenna connected to the receiver.

Method validation

Personnel conducting the radio tracking underwent a series of method validation tests to assess their efficiency at determining bird movement. These tests required each tracker to locate 4 to 6 radio signals from each plot, not knowing whether the transmitters were beacons or attached to birds. Each tracker recorded signal direction, the signal intensity, and whether the signal was moving or stationary. These observations were then compared to the known status of the transmitters. Trackers later used the stationary beacons to assure their receiver was in proper working order before each monitoring session.

Results

Preliminary study

The size of the aviary was sufficient for routine flight by robins. The mesh size and thickness prevented birds from escaping through the netting. The mesh thickness made the mesh visible to birds, and few collisions with the netting were observed.

Transmitter harnesses made of the more rigid, plastic fishing line loops removed feathers from the thighs of birds and required more time to attach. Breast straps were impractical and did not attach well. The breast-mounted harness provided by Wildlife Materials, Inc., although easy and quick to install, appeared to affect bird behavior more than the other types. As a result, the harness selected for the field study was similar in design to that of Rappole and Tipton (1991), but modified by positioning the strands at opposite ends of the transmitter to increase stability. The initial design of the leg-loop harness required hand tying elastic strands, but a subsequent refinement involved pre-tying the loops and embedding the loops in the epoxy base of the transmitter. A loop (tip to tip) size of 62 mm proved to be optimal for attaching transmitters on birds the size of robins, blue jays, and brown thrashers. Elastic stretch accommodated birds which ranged in weight from about 70 to 110 g.

Holohil Systems, LTD transmitters were smallest in size, averaging 1.4 g, and 160 were purchased for the study. Because of production schedules, Holohil Systems was not able to produce all transmitters for the field study. Transmitters from Wildlife Materials were larger at about 4 g, while those provided by ATS, Inc. weighed 4.1 g. Wildlife Materials, Inc. was able to produce a 1.5-g transmitter at a later date, and they supplied 480 of the 640 transmitters required for the study.

Field study

Survival monitoring

A total of 560 songbirds (497 American robins, 46 blue jays, and 17 brown thrashers) were captured in mist nets, radio tagged, and monitored. We detected 500 (89%) of these birds to be alive and present on study plots just prior to pesticide application. Telemetry data collected after the application showed 423 of the 500 (85%) survived until at least 7 d (Tables 8-1 and 8-2). The survival proportion on pesticide-treated plots ($\bar{x}$= 89%) was not less than that on control plots (80%), indicating that the insecticide application had no detectable impact on survival of the focal species.

Tracking validation

Field personnel were able to correctly discern transmitters on free-ranging birds from stationary beacons with a high degree of accuracy (Table 8-3). Stationary beacons were identified correctly 93.2% of the time. Factors such as wind moving the antenna on trans-

Table 8-1   *Number of birds radio tagged, number alive pre- and posttreatment, and proportion of birds which survived on insecticide-treated (Tr) and control (C) plots*

| Golf course name | Treatment type | # birds radio tagged | # birds alive | | Known survivors |
			Pre-[a]	Post-[b]	
Ohio State	Tr	38	34	21	0.61
	C	35	24	5	0.21
Hickory	Tr	40	34	29	0.85
Hills	C	35	24	18	0.75
Columbus	Tr	36	32	32	1.00
	C	37	32	31	0.96
Oakhurst	Tr	32	31	29	0.93
	C	40	39	34	0.87
Blackhawk	Tr	24	22	22	1.00
	C	31	29	24	0.82
Champions	Tr	37	36	34	0.94
	C	29	29	25	0.85
Brookside	Tr	49	43	41	0.95
	C	33	33	27	0.84
Minerva	Tr	33	32	27	0.84
Lakes	C	31	26	23	0.88

[a] Pre-number of birds known to be alive 0 to 3 d before insecticide application
[b] Post-number of birds known to be alive 7 or more d after application

Table 8-2   *Number of birds marked, alive pre- and posttreatment, and percentage of survival by species on insecticide-treated and control plots*

| Species | Control | | | | Treatment | | |
	# marked	# alive pre[a]	# alive post[b]	Percentage survival	# marked	# alive pre	# alive post
American robin	248	218	174	80	249	227	202
Blue jay	15	13	11	85	31	30	28
Brown thrasher	8	5	3	60	9	7	5
Totals	271	236	188	80	289	264	235

[a] # of birds known to be alive 0 to 3 d prior to application.
[b] # of birds known to be alive 7 or more d after application.

Table 8-3 Results of bird movement assessment using radiotelemetry

Golf course	n^a	$\bar{x}$ Deviation degrees[b]	SD	$\bar{x}$% Movement assessment[c]	SD
Hickory Hills	5	9.0	5.3	95.5	11.2
Ohio State	4	6.4	6.4	100.0	0.0
Oakhurst	5	13.0	6.3	95.0	5.1
Columbus	4	21.0	2.9	96.5	7.0
Champions	4	15.5	0.6	92.8	14.5
Blackhawk	4	10.8	4.7	71.9	12.4
Minerva	5	11.6	6.8	97.8	4.9
Brookside	5	9.7	1.9	96.6	7.6
Means		12.1	4.4	93.2	7.8

[a] Number of observers
[b] Average deviation from correct location
[c] Percentage of time transmitter correctly identified birds as moving or stationary

mitters and positioning a transmitter in a tree affected the signal intensity and pulse consistency.

Discussion

Bird survival varied among study sites. The Ohio State University Golf Course demonstrated lower survival than the other 7 golf courses. Factors possibly contributing to this include the time at which birds were netted and radio tagged (late April) and habitat differences. Robins may not have established well-defined territories yet. This golf course was much larger than the other 7 and contained proportionately less nesting habitat within a given plot. Therefore birds may have been using the plots for feeding, but nesting off site. Once nesting began off site, the birds traveled less distance to feed, thus frequenting the golf course less often.

Although there was variation in the survival data, especially in the early data that was unrelated to pesticide use, radiotelemetry was effective in monitoring survival of birds exposed to an application of the pesticide. Higher proportions of marked birds were relocated after the application using telemetry than using other methods (i.e., colored leg bands and patagial tags). Such an increase in monitoring efficiency will greatly improve a researcher's ability to detect pesticide impacts when they occur.

References

Cochran WW. 1980. Wildlife telemetry. In: Wildlife management techniques manual. Washington DC: Wildlife Society. p 507–520.

Mech LD. 1983. Handbook of animal radio tracking. Minneapolis MN: University of Minnesota.

Odom RR, Rappole J, Evans J, Charbonneau D, Palmer D. 1982. Red-cockaded woodpecker relocation experiment in coastal Georgia. *Wildl Soc Bull* 10:197–203.

Raim A. 1978. A radio transmitter attachment for small passerines. *Bird-Banding* 49:326–332.

Rappole JH, Tipton AR. 1991. New harness design for attachment of radio transmitters to small passerines. *J Field Ornithol* 62(3):335–337.

Chapter 9

Use of radiotelemetry to investigate postfledging survival of European starlings following an agricultural spraying of methyl parathion

Michael L. Whitten, Brad T. Marden, Ronald J. Kendall, and Larry W. Brewer

Radiotelemetry was used to investigate the effect of a single aerial application of the organophosphate (OP) insecticide methyl parathion on the postfledging survival of European starlings (*Sturnus vulgaris*). Methyl parathion in a xylene/water solution was applied at 1.1 kg a.i./ha to 2 grass fields. Two similar fields were treated with the xylene/water solution and served as control sites. Nestling starlings were present in nest boxes on all fields and were 8 to 13 d of age at the time of application. At 16 d of age, randomly selected nestlings were equipped with radio transmitters and located daily through radiotelemetry for up to 19 d after fledging. The postfledging survival rate (92%) on the control fields was significantly greater ($P = 0.028$) than the postfledging survival rate (35%) on the treatment fields. These results suggest that an agricultural application of an OP insecticide can result in significant impacts on postfledging survival of altricial birds. The telemetry monitoring used in this study was a very useful tool for rigorously documenting postfledging survival.

Organophosphate compounds represent the largest group of insecticides produced (Smith 1987). Organophosphate insecticides are extensively used in agricultural areas in which numerous species of birds are known to forage, roost, and nest. Applications of OP insecticides often occur during times of peak abundance of birds in agricultural areas as well as during the reproductive season. Many OPs are highly toxic to wildlife and have been responsible for the deaths of large numbers of wild birds (Grue et al. 1983; Smith 1987; Blus et al. 1989). Organophosphate insecticides have also been responsible for several behavioral and sublethal physiological impairments in birds. Lethargy, weight loss, and reduced growth rates have been reported in nestlings following exposure to OPs (Powell and Gray 1980; Grue and Shipley 1981). Hypothermia, increased mortality from predation, and reduced care and feeding of nestlings have been seen in adult birds dosed with various OPs (Grue et al. 1982; White et al. 1983; Galindo et al. 1984; Rattner and Franson 1984). Organophosphate insecticides are toxic primarily due to the ability of their oxon metabolites to inhibit cholinesterase enzymes, causing disruption of nerve function (Smith 1987). Symptoms of OP toxicity include excessive salivation, muscular atrophy, diarrhea, labored breathing, and convulsions. Death usually results from asphyxiation (Murphy 1986).

Most studies of the effects of pesticides on birds have involved controlled dosing of a limited number of pen-reared species. The results from laboratory testing of wildlife have then been extrapolated to other species in the environment. However, there is a paucity of well-documented field studies available for assessing the accuracy of laboratory predictions of pesticide impacts on birds under typical agricultural settings. Because of the extensive usage of OP insecticides and the need for more environmentally realistic exposure conditions, we investigated the effect of a typical agricultural spraying of the OP insecticide methyl parathion (*O,O*-dimethyl *O-p*-nitrophenyl phosphorothioate) on the postfledging survival of European starlings (*Sturnus vulgaris*).

Methyl parathion was chosen as a test chemical because of its widespread use throughout North America at the time of the study. Methyl parathion is a broad spectrum insecticide used primarily on cotton, as well as on soybeans, vegetables, and fruit (Smith 1987). In western Washington, a common application rate for methyl parathion is 0.7 to 1.0 kg active ingredient/ha (kg a.i./ha) for the control of aphids (*Acyrthosiphon* spp.) in crops such as peas (*Pisum* spp.). The rate can be as high as 3 to 4 kg a.i./ha for control of insects in crops such as cotton (*Gossypium* spp.).

Materials and methods

During 18–22 March 1988, 60 wooden nest boxes, patterned after Kessel (1957), were placed on each of 4 agricultural fields located near Ferndale, Washington, and used for hay production or pasture. Each field was 16.2 ha in size and was separated from the other fields by a minimum of 800 m. The predominant plant species on the fields were clover (*Trifolium* spp.), timothy (*Phleum pratense*), and orchard grass (*Dactylis glomerata*). No pesticides had been used on any field for at least 1 y prior to the study. We checked the contents of each nest box at least once every 2 d during the April through June breeding season. Behavioral observations of adult starlings utilizing the nest boxes were conducted intermittently (approximately every 3 d) throughout the study to determine if the starlings were foraging within the study fields. During observation, bird movements were followed as they left the next box and foraged in surrounding areas.

Methyl parathion 4E (USEPA Registration No. 34704-10) in a xylene/water solution was aerially applied at 1.1 kg a.i./ha to 2 fields. The 2 other fields were treated with the xylene/water solution without methyl parathion and served as control sites. The study sites were sprayed on 15 May 1988. The control fields were sprayed immediately prior to the methyl parathion treated fields. Nestling starlings were present in nest boxes on all fields and were 8 to 13 d of age at the time of application.

After pesticide application, radio transmitters were attached to randomly chosen 16 d-old nestlings at each site. Seven nestlings were equipped with transmitters on treatment site #1 and on each control site; 8 nestlings were equipped with transmitters on treatment site #2, for a total of 29 transmitter-equipped birds (15 treatment field nestlings and 14 control field nestlings). At the time of transmitter attachment, pesola scales were used to determine the body mass to the nearest gram of the transmitter-equipped nestlings.

Transmitters were mounted in a backpack fashion on the bird, with attachment threads placed over and under the base of each wing and tied together at the breast. A small drop of waterproof glue covered the knot. Weight of the entire system was 1.3 g. Portable radio receivers were used to locate the fledglings once each day throughout the life of the transmitter batteries (2½ to 3 weeks). The signal reception range varied directly with the height above ground of the bird. Generally, a signal could be received when within 1 km of the bird.

Survival rates of the transmitter-equipped fledglings were calculated using the Kaplan-Meier procedure as presented by Pollock et al. (1989). This procedure estimates the finite survival rate for each monitoring period (day, week, month, etc.) and for the entire study period. The survival function ($S[t]$) is the probability of an individual surviving t units of time from the beginning of the study. The variances of the survival functions were calculated following Cox and Oakes (1984).

The Kaplan-Meier procedure allows the staggered entry of birds during the study. The procedure also allows the censoring of data from birds for which actual fate was not determined due to radio failure or loss of contact with the radio signal.

We used 1-tailed z tests as described by Pollock et al. (1989) to compare differences in survival rate among treatments. We applied the Kaplan-Meier procedure to the group of transmitter-equipped birds on each of the 2 treated and 2 control sites.

One-way analysis of variance (ANOVA) was used to compare body weights of nestlings at 16 d of age. The significance level was $P = 0.05$.

Soil invertebrates (potential starling food items) were collected from a randomly chosen single location on each field for analysis of pesticide residue. Invertebrates were manually removed from the soil (to a depth of 5 cm) and from the accompanying vegetation within a 929 cm^2 (1 ft^2) quadrat. Sampling was conducted on day 2 prespray and days 2, 4, 8, and 16 postspray. Additionally, cranefly larvae (*Tipula paludosa*) were opportunistically collected from the soil surface at 2 locations within treatment field #2 on day 3 postspray.

For pesticide residue analysis, invertebrate tissue was spiked by injecting a known amount of chlorpyrifos (*O,O,*-diethyl *O*-[3,5,6-trichloro-2-pyridinyl] phosphorothioate) into each sample. The tissue was homogenized in a solution of methylene chloride (CH$_2$Cl$_2$) using a Virtis® mechanical homogenizer. The homogenate was then filtered 3 X through a Buchner funnel prepared with diatomaceous earth and dried Na$_2$SO$_4$. The filtered solution was then concentrated via evaporation in a Kuderna-Danish apparatus. Each sample was evaporated down to 2 ml. The samples were then placed in glass vials and stored at 20°C until analyzed for residues. A Hewlett-Packard 5890 gas chromatograph was used for the analysis of pesticide residues. A flame photometric detector was used to detect OP residues.

Results

Analysis of survival revealed no difference within treatment types ($P > 0.05$). Therefore, the 2 control fields were pooled, as were the 2 treatment fields. The postfledging survival rate of starlings on the control sites was significantly greater ($P = 0.028$) than the postfledging survival rate of starlings on the treatment sites (Table 9-1). One fledgling from the control group and 7 fledglings from the treatment group were known to have died during the 19-d monitoring period (Table 9-2).

On the control sites, 1 fledgling was never located after its departure from the nest and 1 was censored (loss of signal) on day 12 of the telemetry study. Seven starlings were censored on days 16 through 19 during the period corresponding to expected battery expiration. The other 4 control group birds were alive at study termination.

Table 9-1   Kaplan-Meier survival estimates ($\hat{S}$) for transmitter-equipped starlings fledged from control and methyl parathion treated sites, and test of the hypothesis of no difference between groups

Site	$\hat{S}$	var $\hat{S}$	Test statistic	
			z	P[a]
Control	0.917	0.010		
Treatment	0.347	0.079		
-	-	-	1.91	0.028

[a] One-tailed z tests

On the treatment sites, 2 fledglings were never located after fledging and 1 fledgling was censored on each of days 9, 10, and 14. One treatment bird was alive at study termination, and the remaining 2 fledglings were censored on days 16 and 18, respectively. Only 1 of the 8 confirmed mortalities was found intact. The carcass of this treatment bird was emaciated with an empty gizzard. Remains of other carcasses were not adequate for necropsy. Portions of 1 carcass were found in a pellet regurgitated from an owl or hawk. The remains of 1 fledgling were found in a burrow 6 to 8 cm below ground. The carcass had been partially consumed, probably by a weasel (*Mustela erminea*). The other recovered carcasses all showed evidence of predation or scavenging. Nestling body weights at 16 d of age did not differ among sites (Table 9-3).

Observations before and after pesticide application indicated that adult starlings foraged primarily within the treated fields when feeding nestlings. Invertebrates collected at the established sampling locations consisted solely of earthworms (*Lumbricidae*). The earthworms contained methyl parathion or methyl paraoxon for at least 8 d on treatment field #1 and 16 d on treatment field #2 (Table 9-4). Methyl parathion residues in earthworms ranged from nondetectable (0.025 ppm) to 2.444 ppm. Methyl paraoxon residues ranged from nondetectable to 1.628 ppm. Cranefly larvae collected from treatment field #2 on day 3 postspray contained 0.067 ppm methyl parathion and 0.025 ppm methyl paraoxon

Table 9-2   Survival of 29 starling juveniles fledged from 2 control fields and 2 fields treated with 1 application of methyl parathion (1.1 kg a.i./ha) and monitored by radiotelemetry. Datapoints begin on the date of fledging. Expected period of transmitter battery expiration: 6–9 June

Bird ID#	May										June								
	22	23	24	25	26	27	28	29	30	31	1	2	3	4	5	6	7	8	9
C-1		o	o	o	o	o	o	o	o	o	o	o	o	o	o	o	o	o	o
C-2			o	o	o	o	o	o	o	o	o	o	x	o	o	o	o	o	o
C-3			o	o	o	o	o	o	o	o	o	o	o	x	o	x	x	x	x
C-4				o	o	o	x	o	o	o	o	x	x	x	x	x	x	x	x
C-5				o	o	o	o	o	o	o	o	o	o	o	o	o	o	o	x
C-6			o	o	o	o	o	o	o	o	o	o	o	o	o	o	o	o	o
C-7				x	x	x	x	x	x	x	x	x	x	x	x	x	x	x	x
C-8			o	o	o	o	o	o	o	o	x	o	o	o	o	x	x	x	x
C-9		o	o	o	o	o	o	o	o	o	o	o	o	o	D				
C-10			o	o	o	o	o	o	o	o	o	o	o	o	o	o	o	o	x
C-11				o	o	o	o	o	o	o	o	x	x	o	x	o	o	o	o
C-12				o	o	o	o	o	o	o	o	o	o	o	o	o	o	x	x
C-13							o	o	o	o	o	o	o	x	o	o	x	x	x
C-14						o	o	o	o	o	o	o	o	x	o	o	o	o	x
T-1		x	x	x	x	x	x	x	x	x	x	x	x	x	x	x	x	x	x
T-2		o	o	o	o	o	o	o	o	o	o	o	o	o	o	x	x	x	x
T-3	x	x	x	x	x	x	x	x	x	x	x	x	x	x	D				
T-4				x	x	x	x	x	x	x	x	x	x	x	x	x	x	x	x
T-5								o	o	o	o	o	o	o	o	o	o	o	o
T-6								o	o	D									
T-7								o	o	o	o	D							
T-8				o	o	o	o	o	x	x	x	x	x	x	x	x	x	x	x
T-9			o	o	o	o	x	x	x	x	x	x	x	D					
T-10						o	o	o	o	o	o	o	o	o	o	o	o	x	x
T-11				o	o	o	o	o	o	x	x	x	x	x	x	x	x	x	x
T-12				o	o	o	x	x	x	D									
T-13					o	o	o	o	o	D									
T-14				o	o	o	o	o	o	o	o	o	o	x	x	x	x	x	x
T-15						o	o	o	o	x	D								

o = bird located; x = bird not located; D = bird found dead.
C-1 through C-8 are Control Field 1.
C-9 through C-14 are Control Field 2.
T-1 through T-7 are Treatment Field 1.
T-8 through T-15 are Treatment Field 2.

Table 9-3   Body masses (g) of 16 d-old transmitter-equipped starlings on control and methyl parathion treated sites

	n	0	SD
Control #1	7	78.3	5.5
Control #2	7	80.4	5.8
Treatment #1	7	75.7	5.4
Treatment #2	8	80.5	5.4

Table 9-4   Methyl parathion (MP) and methyl paraoxon (MO) residue (ppm) in earthworms taken from a single location on treatment sites following application (1.1 kg a.i./ha) of methyl parathion

# days postspray	Treatment field #1		Treatment field #2	
	MP	MO	MP	MO
2	0.251	0.327	2.444	1.628
4	Trace	None	0.535	0.018
8	Trace	None	0.061	0.095
16	None	None	0.041	0.061

Detection limit = 0.025 ppm (wet weight)

at one location and 0.087 ppm methyl parathion at the second location. No methyl parathion or methyl paraoxon was detected on either of the control fields.

Discussion

The Kaplan-Meier procedure assumes that the censoring mechanism is not related to an animal's fate. Possible violations could result from an animal's emigrating because it is stronger than its companions or from the transmitter's being damaged or destroyed through predation of the bird. The possibility of a live bird carrying a defective transmitter would be expected to be the same among treatment and control groups.

The probability of a live bird traveling beyond the range of signal reception, at least in the early postfledging period, is low. Starling fledglings usually do not travel long distances in the first few days after leaving the nest (Feare 1984). Furthermore, intensive efforts were made daily to locate all birds. Three hand-held receivers were used and searches were made on foot in the vicinity of each site and by automobile on all roads within approximately 5 km of each site or within the same distance of each bird's last known location. On one occasion, an aircraft was used to search for missing (censored) birds. No signals from the censored birds were received. It is unlikely that any of these fledglings would have flown beyond the area covered by the airborne search.

A possible explanation for birds censored prior to the period of expected battery expiration is that the birds were dead and the transmitters or antennas had been damaged or destroyed. The antennas on most recovered carcasses were damaged, resulting in severely limited ranges and angles of reception. Signal reception of censored birds may have been rendered impossible by circumstances of predation.

This experiment suggests that a single aerial application of 1.1 kg a.i./ha methyl parathion in an agricultural setting during the nestling phase resulted in nestling exposure that caused a postfledging mortality rate greater than the postfledging mortality rate of starlings on similar control sites. Clearly, postfledging survival should be investigated in any study of the effects of contaminants on avian reproduction.

Stromborg et al. (1988) observed no decrease in postfledging survival of starlings administered a single oral dose of the OP dicrotophos ([E]-Phosphoric acid 3-[dimethyl amino]-1-methyl-3-oxo-1-propenyl dimethyl ester) at 6.0 mg/kg of body mass when 16 d of age. These authors concluded that the effect of a single administration of an OP was reversible and that depression in brain cholinesterase, behavioral changes, or physiological changes were not sufficient to decrease postfledging survival. However, the exposure scenario employed by these authors may not be environmentally realistic and may underestimate the potential for adverse impacts on fledgling survival. Nestlings being fed food items collected from a treated field probably received a chronic low-level exposure over several days.

Contaminated food was available to starlings for several days following application of the pesticide. Methyl parathion and/or its metabolite methyl paraoxon were present in invertebrate samples on treatment fields for 8 to 16 d after application. The 29 transmitter-equipped birds fledged 7 to 15 d after the application of pesticide. Effects on biochemical systems, development, and behavior may have occurred during the nestling and/or the fledgling period through direct exposure to the pesticide via ingestion or dermal absorption. Direct exposure to an OP could cause perturbations in respiration, locomotion, appetite, growth, thermoregulation, vision, learning, and predator avoidance (Powell and Gray 1980; Grue and Shipley 1981, Grue et al. 1982; White et al. 1983; Galindo et al. 1984; Rattner and Franson 1984; Karczmar 1984). Indirect effects, through decreased attentiveness of adults (Grue et al. 1982) or via reduced food availability, may also have diminished the fitness of nestlings or fledglings. There was no evidence of differences in maturity of fledglings among sites, based upon similar body weights of the nestlings at 16 d of age. Although we did not determine which factors were most influential in reducing postfledging survival, it appears that agricultural spraying of an OP caused substantial impacts on altricial birds extending into the postfledging period.

At fledging, altricial birds are suddenly exposed to an environment where their parents can offer only limited protection. The fledglings must quickly learn how to feed themselves and to avoid potential danger. Because the early postfledging period is characteristically one of the most hazardous times in the life of an altricial bird (Feare 1984), it is not surprising that additional stressors, such as pesticide exposure, may significantly increase mortality.

Summary

The effects of a single application of methyl parathion apparently lasted at least 2 to 3 weeks, causing increased mortality in the postfledging period. It is unknown whether the greater mortality of treatment birds was the result of direct OP effects during the postfledging period, physiological impairment during development, reduced food availability, decreased attentiveness of adults, or behavioral modifications caused by exposure during the nestling phase. Nevertheless, the reduced survival of fledglings observed in this study strongly suggests that the postfledging period is a very sensitive time period to investigate in field studies of pesticide impacts on birds. The telemetry monitoring used in this study was a very useful tool through which postfledging survival was rigorously documented.

Acknowledgments - The authors thank Pim Kosalwat for reviewing the manuscript. This work was supported by the USEPA (Cooperative Agreement CR 813662).

References

Blus LJ, Staley CS, Henny CJ, Pendleton GW, Craig TH, Craig EH, Halford DK. 1989. Effects of organophosphorus insecticides on sage grouse in southeastern Idaho. *J Wildl Manage* 53(4):1139–1146.

Cox DR, Oakes D. 1984. Analysis of survival data. London: Chapman and Hall. 201 p.

Feare C. 1984. The starling. New York: Oxford Univ. 315 p.

Galindo JC, Kendall RJ, Driver CJ, Lacher Jr TE. 1984. The effect of methyl parathion on susceptibility of bobwhite quail to domestic cat predation. *Behav Neural Biol* 43:21–36.

Grue CE, Fleming WJ, Busby DG, Hill EF. 1983. Assessing hazards of organophosphate pesticides to wildlife. *Trans N Am Wildl Nat Resour Conf* 48:200–220.

Grue CE, Powell GNV, McChesney MJ. 1982. Care of nestlings by wild female starlings exposed to an organophosphate pesticide. *J Appl Ecol* 19:327–335.

Grue CE, Shipley BK. 1981. Interpreting population estimates of birds following pesticide application behavior of male starlings exposed to an organophosphate pesticide. *Stud Avian Biol* 6:292–296.

Karczmar AG. 1984. Acute and long lasting actions of organophosphorus agents. *Fundam Appl Toxicol* 4:S1–S7.

Kessell B. 1957. A study of the breeding biology of the European starling (*Sturnus vulgaris* L.) in North America. *Am Midl Nat* 58:257–331.

Murphy SD. 1986. Toxic effects of pesticides. In: Klaassen CD, Amdur MO, Doull JD, editors. Casarett and Doull=s toxicologyCthe basic science of poisons, 3rd ed. New York: Macmillan. p 519–581.

Pollock KH, Winterstein SR, Bunck CM, Curtis PD. 1989. Survival analysis in telemetry studies: the staggered entry design. *J Wildl Manage* 53:7–15.

Powell GVN, Gray D. 1980. Dosing free living nestling starlings with an organophosphate pesticide, famphur. *J Wildl Manage* 44:918–921.

Rattner BA, Franson JC. 1984. Methyl parathion and fenvalerate toxicity in American kestrels: acute physiological responses and effects of cold. *Can J Physiol Pharmacol* 62:787–792.

Smith GJ. 1987. Pesticide use and toxicology in relation to wildlife: organophosphate and carbamate compounds. U.S. Fish Wildl. Serv. Resour. Publ. 170, Washington, DC.

Stromborg KL, Grue CE, Nichols JD, Hepp GR, Hines JE, Borne HC. 1988. Postfledging survival of European starlings exposed as nestlings to an organophosphate insecticide. *Ecology* 69:590–601.

White DH, Mitchell GA, Hill EF. 1983. Parathion alters incubation behavior of laughing gulls. *Bull Environ Contam Toxicol* 31:93–97.

Chapter 10

Considerations for planning terrestrial field studies

Lisa M. Ganio

In planning a terrestrial field study, each component of the study should be considered in the context of all other components. There are close connections between the statement of the research question, study design, execution of the study, and final conclusions. These components and their interrelationships are discussed in the context of field studies designed to evaluate effects of pesticides on avian species. Formulation of the research question and design component are emphasized; and attention is given to the topics of randomization, replication, and error control. References for the application of these topics are provided.

Laboratory experiments have been used to help assess the effects of pesticides on avian species (Bennett and Ganio 1991), but the relationship between these results and effects on free-ranging bird populations or individuals exposed to pesticides in the field has not been fully addressed. Pesticide effects in the field can be difficult to duplicate in laboratories. For example, a pesticide may not have a direct impact on birds, but the pesticide may alter some aspect of the ecosystem which, in turn, affects avian species. Field studies can be used to quantify this type of effect, but they are expensive and time-consuming. It is imperative that field studies are carefully planned so that cost-effective and useful information is obtained.

Figure 10-1 depicts the 5 components of a field study: the research question, design of the study, plans for the execution and subsequent analysis of the data collected, and conclusions that are drawn. The figure illustrates the interconnections between each component. The research question is a specific statement that defines the ensuing research. The study design includes the location and definition of the study sites, definition of treatments and randomization plans, descriptions of control measures for sources of variability, and specific statements defining the responses that will be measured. The study design should also define experimental units and replicates if the study is a manipulative experiment or define the sampling plan if the study is an observational study. The execution involves the allocation of time and resources required to carry out the study. Data analysis considerations involve methods, computer software and hardware, and skills required for data analysis. It is also important to consider what kinds of inference might result from the analysis and how these inferences can be used to formulate the final conclusions relevant to the ecological aspects of the study. Planning a field study involves consideration of the potential choices for each component in the context of all of the other components. The final plan should integrate the research question and the constraints of the available resources to provide the best answer to the question based on the most efficient use of the resources.

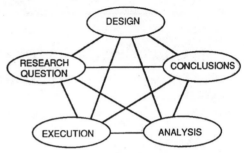

Figure 10-1   Components of a terrestrial field study and their relationships

The purposes of this chapter are to 1) discuss the formulation of the research question and its relationship to the other components, 2) relate the design component to the other components and discuss important aspects of the design of a terrestrial field study, and 3) discuss the links between statistical and biological significance and the final conclusions.

The following quotes by Finney (1978) summarize the major points of this chapter:

> No ingenuity of statistical analysis can force a badly designed experiment to yield evidence on points neglected in the designing.

> The close connection between the design of an experiment and the analysis of the results, however, insures that questions of optimal design cannot be separated from those of statistical analysis.

That is to say, a well-planned study will yield information even if problems occur in the execution or analysis. But a poorly planned study will always be subject to criticism. In addition, the choice of the optimal design depends on the research question and the planned analysis of the data.

Research question formulation

Before any experimental protocols, data collection methods, or data analyses can be proposed, the specific objectives of the study must be defined. An optimal study design for the detection of effects on individual birds may not be the optimal design for detecting an effect on an avian population. Unless the objectives are clearly stated, the field study cannot be made as efficient as possible.

It is important to carefully consider the biological group about which conclusions will be made when defining the research question. This group could be a collection of individual birds at a specific point in time and space. But in a pesticide field study, this group is more likely to be populations of birds that exist across a landscape. The identification of this biological group defines the statistical population from which samples will be taken during the field study. For example, the field study would collect samples from a variety of avian populations when conclusions are desired for multiple populations. Then inference could be made to populations of the types sampled. It is often true that only 1 avian population can be sampled, and it is assumed either explicitly or implicitly that the sampled population was representative of the range of populations of interest. The extent to which this assumption can be made will rest on scientific judgment (Cochran 1977), and it is important to document supplemental information that is used to make such assumptions.

A second consideration in defining the research question is whether the study will be designed to test a statistical hypothesis of a treatment effect (e.g., the pesticide has no effect on hatchling weight), designed to quantify the level of effect (e.g., the pesticide application reduces hatchling body weight by 10%), or designed to quantify differences between the treatment levels (e.g., doubling the application rate decreases hatchling weight by 20%). These distinctions should be included in the research question and will help define the statistical hypotheses or estimation problems in the analysis phase of the study.

The statement of the research question contains specific information about the objectives of the study, the data that will be collected, and the biological group for which conclusions will be drawn. For example, "The goal of this study is to quantify the effect of pesticides on birds" is a statement of general goals, but it does not provide a clear research question. The word "quantify," for example, does not pose the problem in a specific estimation or a hypothesis testing framework. The responses that will be measured in the study are not mentioned. Although birds are mentioned as the group of interest, it is probably not biologically realistic to sample from the population of all birds and to make inference back to that large a group. The researcher probably had a much smaller group in mind.

A better-phrased research question is, "This study will compare the overwintering survival of ring-necked pheasants in Illinois when pesticide A has been applied to corn fields at a rate of 2 pounds per acre every 6 weeks during the preceding growing season as opposed to when the pesticide has not been applied." It is clear that the objective will be determining overwintering survival rates of pheasants, that the biological groups of interest are populations of pheasants in Illinois, and that the study will compare survival between treated and untreated fields. Much of the information required to develop an efficient design is contained in that statement.

When defining the responses that will be measured in the field study, it is important to ask whether the data obtained from the study will be adequate to answer the research question. Understanding the data analysis methods that can be used is necessary to answer this question. In some cases it may be discovered that resources are being wasted by collecting more data than is required. In other cases it may become clear that additional responses should be recorded or a more intense data collection scheme should be used.

A well-stated research question provides clear direction to researchers who must design, execute, analyze, and draw conclusions from a study. Without a clear research question, the study design may be inefficient, resources can be wasted by collecting extraneous data, data analysis can become confused, and conclusions that are not entirely helpful may be obtained. Clear statements of research questions will point to areas where additional study is needed as well as help coordinate research among collaborating scientists and explain the purpose of the research to the nonscientific community.

Planning the study design

Careful consideration of the research question, the available resources and physical limitations of the study, the data analysis, and the final inference that is desired can result in an efficient study that provides a useful result. The design of a field study should define the choice of field sites and the pesticide application rates, the process by which the application rates are assigned to the sites, measures taken to control known and unknown sources of variation, and specific response variables that will be measured. This section discusses issues that must be considered in selecting choices for each of these parts of the design.

Setting the limits of the study

Selection of each aspect of the design from a pool of potential choices should be based on information from the other components of the study. A good design will be based on careful consideration of all components. The research question may define the data to collect and define the location of the sites used. For example, researchers studying pheasant survival in Iowa must collect data on the survival of pheasants, and the sites used should primarily be in Iowa. The number of sites used and the measures used to control sources of variability depend on the money, time, and manpower available to the study. The decision to use a blocking factor to control climatic variation may not be made until after the general location of the sites is chosen. The method by which application rates are assigned to sites may also depend on the sites chosen. For example, if all the sites in the study are physically close, then the random assignment of treatments among sites may be possible. However, if sites are scattered in a variety of states, the random assignment might have to be carried out state by state.

A cost-benefit analysis is a formalization of this idea. The costs associated with materials, execution, and analysis can be estimated for various designs and weighed against the potential outcomes of the research. This information can be used to select designs that provide a balance between costs and information. Examples of cost functions in the context of general sampling problems have been given by Cochran (1977), while Skalski and Robson (1992) provided examples of the impact of cost on design considerations for field studies that involve mark/recapture or radio tagging. Even when a formal cost analysis is not conducted, it is important to consider whether or not all of the resources necessary to carry out a design and its analysis (e.g., personnel, radio collars, vehicles, computers, and computer software) are available and functioning within the required timeframe. If the resources are not adequate to meet the study objectives, the study may need to be redesigned or aborted.

Observational versus experimental studies and the randomization process

Eberhardt and Thomas (1991) distinguished between research in which the investigator controls the circumstances of the study and studies that involve passive observation. They suggested that, "The major error made in many contemporary analyses is that of confusing control of events (treatments) with control of the observational process." Control of the assignment of treatments to experimental units is the defining feature of an experimental or manipulative study, and its use affects the conclusions that can be drawn (Cox 1958; Skalski and Robson 1992). If the assignment process is not under the researcher's control, then the study generally is not experimental in nature and should be labeled an observational study or a sample survey (Skalski and Robson 1992). Skalski and Robson (1992) also distinguish between impact studies which lack randomization and assessment studies which lack replication. In pesticide field studies where concern centers on the effect of the chemical, it is the method by which pesticide applications are assigned to fields that distinguishes experimental from observational studies. If each application is randomly assigned under the control of the researcher to each field, then the study is a manipulative experiment. But if the assignment of pesticide application is made using a process other than random assignment, the study is an observational study.

Two points deserve mention here. First, the nonrandom assignment of pesticide applications to fields is a likely scenario in pesticide field studies because the decision is often controlled by financial rather than scientific concerns. Secondly, the principles of design apply to observational, impact, and assessment studies as well as to manipulative experiments; therefore, the phrase study design, rather than experimental design, is the focus of this chapter.

Random assignment is defined to be the process that assigns to every potential unit a known, usually equal, probability of being selected for any treatment group. Such a method is definable, replicable, and quantifiable. It is contrasted with the method of subjective assignment which often cannot be well-described, quantified, or documented, but the distinction between subjective and random assignment can be unclear. Some of the confusion over the use of this term results from the use of the definition in a nonstatistical setting. The phrase "at random" is often used to mean "without careful choice, aim, or plan" (Neufeldt and Duralnik 1988). While this definition implies a lack of uniformity and replicability, the definition of a random sample, given as "sample selection in which all possible samples have equal probability of selection," implies a repeatable selection process based on uniform selection probabilities. Thus when the phrase "random assignment" is interpreted in an English language context, it is often interpreted as implying nonuniformity, whereas statistically it should imply uniform selection probability. Researchers must be aware that subjective assignment will not substitute for random assignment.

Random assignment guards against identifiable, but uncontrollable, sources of bias and unknown sources of bias by ensuring that each unit is no more likely to be assigned to a given treatment than is any other unit. Thus systematic differences between experimental units due to an uncontrolled factor are unlikely to exist (Skalski and Robson 1992). The use of randomization helps protect against systematic differences between treatment groups that are due to factors other than the assigned treatment. When random assignment is not used, a factor may differ systematically across treated and untreated units, and it may unknowingly be responsible for an observed effect.

When results are obtained that indicate differences between treatments in a manipulative experiment, it is the use of random assignment that allows researchers to conclude that the pesticide application was responsible for the observed effect. It can be concluded that the differences are due to the treatments because theoretically any other effects have been randomized across treatments. However, in an observational study, no precautions are used to minimize effects of unknown factors; therefore, the formal conclusion that can be drawn should be phrased as an association between treatment and effect. Since the presence of randomization in a study has a direct bearing on the wording of conclusions that can be drawn, its use should be carefully considered in the design phase.

In Chapter 4, Cox (1958) pointed out the difficulties in the interpretation of nonrandomized studies and provided examples that illustrate the disadvantages of using methods other than random assignment. He discusses methods for handling problems that arise when the number of units subject to randomization is small.

Replication

Replication of treatments, referred to as "replication," is another important but often confusing design element. When replication is used along with randomization, the variability between the units which receive the same treatment can be estimated and differences between treatment groups are evaluated relative to the size of this estimated experimental error. Cox (1958) defined a replicate as the smallest piece of experimental material that might receive a different treatment, i.e., a replicate receives an independent application of the treatment.

When a researcher does not control the assignment of the treatment to an experimental unit but instead controls the observational process and confuses this with true replication, then pseudoreplication as defined by Hurlbert (1984) can occur. For example, 4 measured subplots within a single field that received 1 pesticide application are not 4 replicates of the treatment. The variability between these subplots (within 1 replicate) can be used to assess the observational process, but not the differences between treatments. In Chapter 8, Cox (1958) discussed the composition of the error term for the comparison of mean values as a function of both the between-treatment and within-treatment variation and analysis methods when 1 type of variation is not measured. Skalski and Robson (1992) suggested methods for estimating variance components and for using this information to optimize the design of the field study in the context of mark/recapture studies.

In some pesticide field studies, replication is not possible or is too costly to use. When a lack of replication is foreseen in the planning stage, it may be possible to use designs or treatment structures in which experimental error can still be estimated. Milliken and Johnson (1989) provided case studies and examples of designs and analysis methods for studies lacking true replication. When independent estimates of variability cannot be obtained, however, the application of the results to situations beyond those in the study must be limited.

Controlling known sources of variation

The study design can also include measures to control predictable and known sources of variation. The use of control sites, blocking factors, and covariate measurements are methods that can help control variability in responses, but their use must be carefully planned in the context of the entire field study.

Control sites can be used in a variety of ways in the context of field studies. Control sites can be similar in all respects to sites that receive the pesticide except that they do not receive any application. These sites are included to assess the ecological effects of the pesticide above and beyond naturally occurring events. Or, control sites can receive a pesticide that is thought to result in minimal or acceptable ecological effects. The responses obtained from these control sites are considered representative of acceptable pesticide effects and are contrasted with effects from the pesticide of concern. The third type of control site receives an application that is similar in all respects to the pesticide application except that no chemical and only a neutral carrier is applied. This method allows any effects that are due solely to application technique and not to the pesticide to be accounted for in the analysis. Control sites can also be used to account for naturally occurring temporal variability in the analysis. When responses are monitored on the same fields before and after the pesticide application and control sites are not used, temporal effects cannot be separated from treatment effects.

Blocking is another method for reducing extraneous variability. A statistical block is a group of units to which treatments will be applied and which are more alike than the units in a different block. For example, in a study to assess decreases in abundance of robins due to pesticide use on golf courses and apple orchards, a set of plots, all of which are on golf courses, may receive independent applications of either the pesticide or a neutral carrier. Another set of plots in apple orchards may receive similar independent applications. Comparisons between pesticide and the neutral-carrier-treated plots are made within the golf course plots and within the apple orchard plots, excluding the variability that would have existed if the comparisons had been made between all pesticide plots and all control plots.

In some cases it might be known that certain quantifiable environmental factors will affect the efficacy of the pesticide (e.g., the air temperature and humidity just prior to a spray event). If these covariates, or concomitant variables, are measured along with the response variables for each experimental unit, they can be used to help explain extraneous variation in the data. The major assumption implied by their use is that the covariate is not influenced by the treatment. Cox (1958) discussed how to use these types of measurements, and Cochran and Cox (1957) discussed what kinds of information can be gleaned from experiments where covariates are affected by treatments.

The use of blocking and covariates are powerful means of reducing variation when their use is justified. However, when an ineffective blocking factor or covariate is used, these procedures can reduce the overall ability to detect real differences between treatments. It is important to be as certain as possible that units within blocks are more alike than units between blocks or that potential covariates do indeed explain some of the variability in the data when they are used. Pilot studies or previous research may help provide some of this information.

As stated above, the execution and analysis of a field study should be considered in its design phase. Once a plan for these phases of the research has been devised, the actual implementation is less subject to mistakes due to poor planning. When unexpected circumstances occur that lead to changes in execution or analysis, alternatives are more likely to have been explored.

There is a myriad of data analysis methods that might be applicable to field studies involving radiotelemetry. The reader is referred to Otis et al. (1978), Burnham et al. (1987), and Pollock et al. (1990), for discussions of mark/recapture models and methods; to Meretsky (1987), Pollock et al. (1989), and White and Garrott (1990) for treatment of data from radiotelemetry studies; and to Skalski and Robson (1992) for a good overview of existing methods.

Linking statistical analysis and biological conclusions

Once the study has been carried out and the data analyzed, the researcher will be able to evaluate the results and draw conclusions about the effects of the pesticide. These conclusions should be made relative to the ecology of the problem and not solely on the basis of a statistical test. The detection of a statistically significant effect should be interpreted relative to its impact on the population in question over a reasonable time period.

Planning how to relate any statistical results to a biological group is critical to obtaining useful results. Researchers should consider how potential statistical results might lead to ecological conclusions that can answer the research question. If the data will not result in conclusions that can be applied to the overall question, then changes in the data that will be collected or execution and analysis may need to be considered. Sometimes this process is obvious. For example, if a treatment resulted in a mortality rate of 80%, it could be inferred that an 80% reduction in an avian population would be an important ecological effect. But when results are not directly linked to a demographic effect, e.g., significant weight loss or changes in activity patterns, the researchers must consider how these effects relate to the ability of the birds or populations of birds to survive, breed, and interact with their environment.

Summary

Every field study will be a unique study of a given system. Although field studies will contain many common elements, there will always be aspects of each that require special consideration and planning. The inclusion of a trained statistician early on and throughout the planning phase can be invaluable in helping to address difficult problems such as a lack of replication or effective methods of variation control. Addressing statistical issues at this stage in the research, rather than after the data are collected, can result in more powerful analyses and stronger conclusions.

Careful planning of a field study is required to make the best use of available resources in providing the most complete answer to the research question. Choices for each component of the study should be evaluated with respect to their effect on all other components. A well-designed study will include components that compliment and relate to each other in logical ways. The planning process will include input from primary investigators, field crews, data analysts, statisticians, and the final users of the research. In addition, potential changes in any component will be considered so that unexpected changes in protocol during execution of the study can be accommodated.

References

Bennett RS, Ganio LM. 1991. Overview of methods for evaluating effects of pesticides on reproduction in birds. USEPA, Environmental Research Laboratory, Corvallis OR. EPA-600/391/048.

Burnham KP, Anderson DR, White GC, Brownie C, Pollock KH. 1987. Design and analysis methods for fish survival experiments based on release-recapture. *Am Fish Soc Monogr* 5.

Cochran WG. 1977. Sampling techniques. 3rd ed. New York: Wiley. 428 p.

Cochran WG, Cox GM. 1957. Experimental designs. New York: Wiley. 611 p.

Cox DR. 1958. Planning of experiments. New York: Wiley. 308 p.

Eberhardt LL, Thomas JM. 1991. Designing environmental field studies. *Ecol Monogr* 6(1)53–73.

Finney DJ. 1978. Statistical method in biological assay. 3rd edition. London: Griffin. 508 p.

Hurlbert SH. 1984. Pseudoreplication and the design of ecological field experiments. *Ecol Monogr* 54(2):187–211.

Meretsky V Jr. 1987. Experimental design and data analysis for telemetry projects: summary of a workshop. *J Raptor Res* 21(4):125–146.

Milliken GA, Johnson DE. 1989. Analysis of messy data, Vol. 2: Nonreplicated experiments. New York: Van Nostrand Reinhold. 199 p.

Neufeldt V, Duralnik DB. 1988. Webster's new world dictionary of American English. New York: Webster's New World. 1574 p.

Otis DL, Burnham KP, White GC, Anderson DR. 1978. Statistical inference from capture data on closed animal populations. *Wildl Monogr* No. 62.

Pollock KH, Nichols JD, Brownie C, Hines JE. 1990. Statistical inference for capture-recapture experiments. *Wildl Monogr* No. 107.

Pollock KH, Winterstein SR, Bunck CM, Curtis PD. 1989. Survival analysis in telemetry studies: the staggered entry design. *J Wildl Manage* 53:7–15.

Skalski JR, Robson DS. 1992. Techniques for wildlife investigations: design and analysis of capture data. San Diego CA: Academic. 237 p.

White GC, Garrott RA. 1990. Analysis of wildlife radio-tracking data. San Diego CA: Academic. 383 p.

Chapter 11

Analysis of radiotelemetry survival data in an avian field study

Lyman L. McDonald, David L. Fischer, and Paul A. Toll

A paired-plot manipulative experimental design was used for a screening study of the effect on birds of an experimental compound when used on golf courses for insect control. For each of 8 pairs of plots, 1 member of the pair was selected at random to receive the compound. The experimental procedures for capture of birds, radio tagging, and tracking of tagged individuals followed uniform procedures within replicate pairs of sites. Define

   μ_c = mean of survival rates of tagged birds known to be present on control plots during a certain posttreatment period, and

   μ_t = mean of survival rates of tagged birds known to be present on treated plots during the same posttreatment period of time.

We consider statistical procedures for testing the hypothesis that pesticide application reduces the mean survival rate by 20% or more compared to the controls, i.e., we test the hypothesis

   H_o: $\mu_t \le (0.8)\,\mu_c$

against the alternative

   H_a: $\mu_t > (0.8)\,\mu_c$.

We also investigate the power of the paired t-test to reject the hypothesis that the pesticide has an effect on survival rates using model-based Monte Carlo computer simulation over anticipated ranges of the population parameters and sample sizes.

We report on the analysis of data arising in a field study of the acute hazard to birds caused by application of an experimental insecticide (the "treatment") to golf courses. The active ingredient of the treatment is formulated on bentonite clay granules about 2.2 mm in diameter and weighing on average 0.44 mg. The proposed application of this formulation is on turf grass on sites such as golf courses, sod farms, and home lawns to control white grubs and other soil insects primarily in the Northeast and North Central United States. The granular insecticide was applied by broadcast at the maximum rate in accordance with use directions on the proposed label. Potential routes of exposure to birds include direct ingestion of granules, ingestion of residues in or on avian foods or surface water, and dermal absorption.

The primary objective of this chapter is quantification of the impact of the treatment on survival of a radio-tagged sample of birds potentially at risk. Golf courses were chosen over sod farms and other turf grass sites for the study because golf courses contain a diversity of woodland, wooded parkland, and shrubland interspersed with turf grasses. These habitats were expected to contain more species and numbers of birds than would be found at more open sod farms. If the treatment is found to be acceptable on golf courses, then it may be presumed to also be acceptable at other application sites.

The second objective is to investigate the power of the proposed statistical procedures to reject the null hypothesis that the treatment does have an effect on survival rates. Power values are approximated by Monte Carlo computer simulation over anticipated ranges of the population parameters and number of replicated study sites.

Methods

Golf course study

The study was conducted at 8 golf courses during the avian breeding season (mid-April through mid-July). The breeding season was selected for study because avian populations are at their most sedentary and the treatment was commonly made at this time. Three focal species were selected to be outfitted with radio transmitters. These species were among those that 1) are at relatively high risk, 2) may be easily monitored, 3) are resident at golf courses, 4) occupy fairly small home ranges so that they are likely to stay mainly within the plot where captured (treatment or control plot), and 5) are abundant enough to permit a reasonably precise estimate of survival rates. These species were the brown thrasher (*Toxostoma rufum*), the blue jay (*Cyanocitta cristata*), and the American robin (*Turdus migratorius*). The 8 golf courses were selected based on presence of habitat types that should contain an appropriately large population of the focal species.

Two study plots, each consisting of 5 to 9 golf holes, were established on each course. Plots were chosen such that they were in different regions of the course in an attempt to minimize problems with movement of the focal species between the 2 plots. One of the plots, chosen at random, was treated. The other (control) plot and the rest of the course were not treated during the study period. There was no use of other highly toxic compounds during the study, and all experimental procedures were uniform within replicate pairs of sites.

Approximately 3 to 21 d prior to treatment, American robins, brown thrashers, and blue jays were captured and outfitted with radio transmitters. Birds were captured in mist nets and affixed with an aluminum U.S. Fish and Wildlife Service leg band (a unique combination of colored plastic leg bands, a corresponding combination of colored patagial tags, and a radio transmitter) using techniques described by Rappole and Tipton (1991). A maximum of 80 birds were to be radio tagged per course. The status of birds (alive, dead, not present) receiving radio transmitters was determined at least every 5 d, including 3 d pretreatment (day −3), treatment day (day 0), 3 d posttreatment (day +3), 7 d posttreatment (day +7), and 10 d posttreatment (day +10). Locations of radio-tagged individuals detected were recorded.

On treatment day, the granular insecticide was applied by broadcast at the maximum rate in accordance with use directions on the proposed label. The application was made by personnel who normally apply granular pesticide treatments at each course following their standard procedures so long as the proposed label was not violated. However, the application was not "watered in" immediately after application to reduce avian risk because the proposed label did not require this practice.

In addition, other activities associated with the study, but which were not directly related to the analysis of radiotelemetry survival data, were conducted (e.g., carcass searches, monitoring of environmental residues, etc.).

Statistical analysis methods

Define

μ_c = mean survival index of tagged birds known to be present on control plots during a certain posttreatment period, and

μ_t = mean survival index of tagged birds known to be present on treated plots during the same posttreatment period of time.

We test the hypothesis that pesticide application reduces the mean survival index by 20% or more compared to the controls, i.e., we test the hypothesis

$$H_o : \mu_t \leq (0.8)m_c$$

against the alternative

$$H_a : \mu_t > (0.8)m_c.$$

To estimate an index on the mean survival rate for each site, we pooled the results of the 3 focal species. We computed the number of individuals known to be present on day -3 or day 0 and the proportion of those individuals known to still be present on day $+7$. Birds were counted as being present on day $+7$ if they were confirmed present on day $+7$ or day $+10$. Birds on treated plots were assumed to be exposed to the treatment. For the *i*th pair of sites, the proportion of radio-tagged birds which was known to be alive 7 d posttreatment was denoted by the pair (p_{ti}, p_{ci}) where subscripts of t and c denoted the treated and control plots respectively. We conducted standard 1-sided tests on the mean difference of these pairs using the paired t-test and Fisher's permutation test. Results are presented graphically by inverting the paired t-test to obtain the lower limit of a confidence interval on the ratio of the mean survival rates (Miller 1986), as proposed by BH Stanley and JD Wetherington and reported in McDonald et al. (1992). For convenience of the reader, the formula for the paired plot design is reproduced in the appendix following this chapter.

Monte Carlo simulation of power

The survival indices can be viewed as the proportion of successes where the sample size (i.e., the number of birds marked on a given plot) varies from one observation to the next. Using Monte Carlo computer techniques, McDonald et al. (1992) simulated the power of the paired t-statistic to reject the hypothesis of assumed effect in field studies such as those described above. We report selected results from that simulation study.

Using a set of data from a study conducted by Wildlife International, Ltd., Easton, Maryland, McDonald et al. (1992) set the mean and standard deviation of the number N of birds captured and available immediately prior to treatment to $\bar{x} = 26.3$ and $S = 12.7$, respectively. They then generated values $\sqrt{N}$ from a normal distribution with mean 4.97 and the standard deviation 1.26. This transformation is used because in many situations the count data, N, are often non-normal but the distribution of $\sqrt{N}$ (the variance stabilizing transformation of the Poisson distribution) is approximately normal. To check the influence of fewer birds captured and known to be present at the time of application of the treatment, simulations were conducted for means of 18 and 12 for N with standard deviations of 4 and 3, respectively.

McDonald et al. (1992) investigated the performance of the paired t-test across a range of values for the survival index. They simulated the power to reject the hypothesis of an assumed effect with treatment effects ranging from a placebo treatment to substantial reductions in the survival index. The effect of pesticide treatment was simulated by reducing the survival index for control plots by the "treatment effect factor" b = 0.70, 0.75, 0.80, 0.85, 0.90, 0.95, or 1.00 (placebo effect). Expressed in a formula, the mean treatment survival rate was $\mu_t = b(\mu_c)$ for each treatment effect factor. A case of special interest is the power of the tests with b = 1.0, i.e., the power to reject the hypothesis of an assumed effect when one is testing a placebo treatment ($\mu_t = \mu_c$). On the other extreme, b = 0.8 corresponds to the case when the hypothesis of an assumed 20% reduction in the survival rate is true and the simulated power should be approximately equal to the target value of the size of the test, α. We report the results of particular interest for testing the hypothesis of assumed effect on survival rates in the paired plot design. McDonald et al. (1992) report the power for other cases including randomized block designs with blocks of size 3, mortality searches, and the standard null hypothesis of no treatment effect.

McDonald et al. (1992) generated 5,000 replications of the simulated experiments for each combination of

1) mean and standard deviation of the survival index (Table 11-1);
2) treatment effect factor b = 0.70, 0.75, 0.80, 0.85, 0.90, 0.95, and 1.0; and
3) size of experiment (ranging from 6 to 16 pairs of plots).

Their simulation consisted of 8 steps:

1) Generate the pretreatment number of radio-tagged birds for each of the control and treatment plots by squaring and rounding off the results of simulating variates from a normal distribution with mean 4.97 and standard deviation 1.26.

2) To simulate the influence of pairing plots to control sources of variation, the same survival index was used for both members of the pair of plots. Generate a survival

Table 11-1   Mean and 5 possible standard deviations (A, B, C, D, E) of the survival index on control plots used in the simulation studies of power to reject the hypothesis of assumed effect

Survival index on controls		Standard deviations				
		A	B	C	D	E
	0.2	0.02	0.03	0.04	0.05	0.06
	0.4	0.04	0.08	0.12	0.16	0.20
	0.6	0.06	0.09	0.12	0.15	0.18
Mean	0.8	0.04	0.08	0.12	0.16	0.20
	0.9	0.045	0.09	0.135	0.18	0.225
	0.95	0.0475	0.095	0.1425	0.19	0.2375

index for each pair of plots from a truncated normal distribution (so that $0 < s < 1$) with the given mean and standard deviation from Table 11-1.

3) Generate the number of birds relocated posttreatment on a control plot from a binomial distribution with probability of survival equal to the generated survival index for the pair of plots and the number of trials equal to the pretreatment number of radio-tagged birds for that control plot.

4) Calculate modified survival indices for the treated plots by multiplying the control survival indices by b.

5) Simulate the number of birds relocated on the corresponding treated plot with a binomial distribution, this time with a probability of survival equal to b times the survival index generated for the pair of plots and with the treated plot's own generated pretreatment number of radio-tagged individuals.

6) Calculate the observed value of the survival index for each plot as the ratio of the number of individuals relocated posttreatment to the pretreatment number of radio-tagged individuals present.

7) Calculate the appropriate t-statistic for the experiment.

8) Repeat the simulation 5,000 times and compute the power by checking the number of times the t-statistics lead to rejection of the hypothesis at the chosen significance level.

Results

Golf course study

Results for the 3 focal species combined are given in Table 11-2 where it is seen that the mean value for $(p_t - 0.8(p_c))$ is 0.271 with standard error 0.0324. The paired t-statistic for the hypothesis of at least a 20% reduction in the mean survival index to 7 d posttreatment for the 3 focal species combined is t = (0.271/0.0324) = 8.36 (P-value < 0.0001). Following Miller (1986), the lower limit of a 90% 1-sided confidence interval on the ratio (μ_t/μ_c) is 0.968 (Figure 11-1). In other words, one can make the statistical conclusion with 95% confidence that the treatment mean survival index is above 96.8% of the control mean survival index.

Monte Carlo simulation of power

A summary of results from McDonald et al. (1992) is given. In Table 11-3, we report n, the minimum numbers of pairs of plots, needed to achieve at least 80% power to reject the hypothesis of an assumed 20% effect (i.e., $\mu_t = (0.8)\mu_c$) when b = 1.0 (placebo treatment effect), 0.95, 0.90, or 0.85 using tests of size α = 5% or 20%. The actual power level that will be met with the indicated number of pairs is also reported.

Table 11-2   Summary of estimated survival index to 7 d posttreatment for all radio-tagged birds. $N_{.3}$ = number radio-tagged birds present 3 d pretreatment or on treatment day, N_7 = number radio-tagged birds still present 7 days posttreatment, p_7 = proportion radio-tagged birds still present 7 days posttreatment. $_c$ and $_t$ denote control and treatment plots respectively.

Replicate golf course	Control plots			Treated plots				
	$N_{c,-3}$	$N_{c,7}$	$p_{c,7}$	$N_{t,-3}$	$N_{t,7}$	$p_{t,7}$	$0.8(p_{c,7})$	$p_{t,7}-0.8(p_{c,7})$
1	24	18	0.750	34	29	0.853	0.6000	0.2529
2	24	5	0.208	34	21	0.618	0.1667	0.4510
3	39	34	0.872	31	29	0.935	0.6974	0.2380
4	32	31	0.969	32	32	1.000	0.7750	0.2250
5	29	24	0.828	22	22	1.000	0.6621	0.3379
6	29	25	0.862	36	34	0.944	0.6897	0.2548
7	33	28	0.848	43	41	0.953	0.6788	0.2747
8	26	23	0.885	32	27	0.844	0.7077	0.1361
							Mean	0.271306
							Standard deviation	0.091718
							Standard error	0.032427

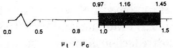

Figure 11-1   Limits of a 90% confidence interval on ratio of mean survival index for treated plots relative to mean survival index for control plots, μ_t/μ_c. Lower limit interval corresponds to a 95% 1-sided confidence interval on ratio.

Another meaningful way to report the power of these experiments is to test the hypothesis of an assumed effect

$$H_o \;:\; \mu_t \leq b(\mu_c)$$

for b = 0.70, 0.75, 0.80, 0.85, 0.90, and 0.95 against

$$H_1 \;:\; \mu_t > b(\mu_t)$$

when in fact $\mu_t = \mu_t$, i.e, when the treatment effect is equivalent to that of a placebo. A summary of these power computations are reported in Table 11-4. Note that row 4 and row 8 of Table 11-3 correspond with row 4 and row 10 of Table 11-4 because the null and alternative hypotheses turn out to be the same in those cases (slight differences exist in the power values because independent simulations were used to generate the 2 tables).

Table 11-3 Minimum number[a] of pairs of plots and power (n; power%) to achieve at least 80% power to reject the hypothesis of assumed 20% effect ($\mu_t = (0.8)m_c$) using tests of size $\alpha = 5\%$ and 20%. Power values reported are for n = 6 when the simulation indicated that at least 80% power is achieved for n $\leq$ 6 because the minimum value on n considered in this simulation was 6. Mean number of radio-tagged birds is $\bar{x} = 26.3$ with standard deviation 12.7

α	Treatment factor b	Mean survival index on control plots					
		0.20	0.40	0.60	0.80	0.90	0.95
5%	0.85	-	-	-	-	-	-
5%	0.90	-	-	-	-	(11;80.3)	(9;81.2)
5%	0.95	-	-	-	(8;80.1)	($\leq$ 6;84.4)	($\leq$ 6;91.0)
5%	1.00	-	-	(11;83.4)	($\leq$ 6;88.0)	($\leq$ 6;95.6)	($\leq$ 6;97.6)
20%	0.85	-	-	-	-	-	(15;80.3)
20%	0.90	-	-	-	(8;81.9)	($\leq$ 6;87.1)	($\leq$ 6;89.8)
20%	0.95	-	(16;80.5)	(8;81.6)	($\leq$ 6;91.9)	($\leq$ 6;97.2)	($\leq$ 6;98.5)
20%	1.00	-	(9;81.0)	($\leq$ 6;87.2)	($\leq$ 6;98.3)	($\leq$ 6;99.4)	($\leq$ 6;99.7)

An entry of - indicates that 80% power was not achieved with 16 pairs, the maximum considered

Discussion

Golf course study

The hypothesis of assumed reduction of at least 20% in the mean survival index relative to the control plots for the 7 d posttreatment is soundly rejected. In fact, this statistical inference is straightforwardly based on the fact that none of the values of p_t are below $0.8(p_c)$. Any of the standard nonparametric tests, including Fisher's permutation test, would lead to the same statistical conclusion. Use of the 1-sided 95% confidence interval on the ratio μ_t/μ_c provides the further statistical conclusion that the ratio is above 96.8%. This statistical inference concerning the mean index on survival strongly supports the conclusion that there is little if any reduction in survival due to the treatment.

Table 11-4   *Minimum number[a] of pairs of plots and power (n, power%) to achieve at least 80% power to reject H_o; $\mu_t \leq b(\mu_c)$ when in fact $\mu_t = \mu_c$ using tests of size α = % or 20%. Power values reported are for n = 6 when the simulation indicated that at least 80% power is achieved for n $\leq$ 6 because the minimum value on n considered in this simulation was 6. Mean number of radio-tagged birds is $\bar{x}$ = 26.3 with standard deviation 12.7*

α	Treatment factor b	Mean survival index on control plots					
		0.20	0.40	0.60	0.80	0.90	0.95
5%	0.95	-	-	-	-	-	-
5%	0.90	-	-	-	(16;81.3)	(9;81.7)	(7;82.5)
5%	0.85	-	-	-	(8;82.2)	(≤6;86.8)	(≤6;92.6)
5%	0.80	-	-	(11;83.1)	(≤6;87.1)	(≤6;95.5)	(≤6;97.8)
5%	0.75	-	(14;81.7)	(7;80.2)	(≤6;95.2)	(≤6;98.6)	(≤6;99.4)
5%	0.70	-	(10;82.1)	(≤6;85.7)	(≤6;98.4)	(≤6;99.4)	(≤6;99.8)
20%	0.95	-	-	-	-	(14;80.9)	(10;82.0)
20%	0.90	-	-	-	(7;81.1)	(≤6;89.6)	(≤6;94.0)
20%	0.85	-	-	(8;81.5)	(≤6;92.8)	(≤6;97.6)	(≤6;98.8)
20%	0.80	-	(9;80.1)	(≤6;87.7)	(≤6;98.0)	(≤6;99.5)	(≤6;99.7)
20%	0.75	(15;80.1)	(≤6;81.3)	(≤6;94.6)	(≤6;99.5)	(≤6;99.9)	(≤6;99.9)
20%	0.70	(10;80.3)	(≤6;89.3)	(≤6;97.8)	(≤6;99.8)	(≤6;99.9)	(≤6;99.9)

An entry of - indicates that 80% power was not achieved with 16 pairs, the maximum considered

One golf course (#2 in Table 11-2) showed an extremely low survival index on the control plot (20.8%) and relatively low survival index on the treated plot (61.8%). This pair of plots provides no evidence of a treatment effect, but apparently an outside factor such as weather may have caused some migration from the plots. If this golf course is removed from the dataset, the resulting paired t-statistic is t = 10.74 with 6 degrees of freedom, and it remains significant at the P ≤ 0.0005 level.

Monte Carlo simulation of power

The power values in Tables 11-3 and 11-4 indicate that good to excellent power to reject the hypothesis of assumed effect exists using the paired plot design if the survival index on control sites is high (e.g., if the study is conducted during the breeding season). If the study reported in this chapter were to be repeated, then we could expect a survival index of about 80% on control plots and about N = 30 birds to be radio tagged on each plot (slightly better than in the simulations). From Table 11-3, the power to reject the hypothesis of at least 20% reduction in the survival index is about 88% for α = 5% and n = 8 if the treatment has no effect (b = 1). With a 5% reduction in the survival rate due to the treatment, the power to reject the hypothesis of at least 20% reduction is about 80% for α = 5% and n = 8. If a test of size α = 20% is used (which may be allowed in certain worst-case studies) (Fite et al. 1988), the power is much better. Therefore, in some circumstances, studies can be designed using the paired plot method that yields good to excellent power to refute the hypothesis of assumed 20% effect on survival with a moderate number of replications.

With low values for the survival index on control plots, as may be expected if many birds are migrating through the region or if many signals are lost from radio-tagged birds, the power to reject the hypothesis of assumed effect is very low. When the mean survival index is 0.20 and the assumed treatment effect factor is 0.80 but, in fact, $\mu_t = \mu_c$, the power is only 0.38 (α = 5%, 16 pairs of plots); however, the power increases to 0.70 (α = 5%, 16 pairs of plots) if the mean survival index is 0.40 (McDonald et al. 1992).

In general, there is not much power lost in dropping from $\bar{x}$ = 26.3 individuals marked per plot to $\bar{x}$ = 18 individuals marked per plot. In fact, the power for an $\bar{x}$ = 18 marked birds per plot did not drop much below 90% of the power when $\bar{x}$ = 26.3 for the same values of the other parameters (McDonald et al. 1992). However, for some cases, power for a mean of $\bar{x}$ = 12 marked birds per plot dropped to slightly below 70% of the power for $\bar{x}$ = 26.3 for the same values of the other parameters. For preliminary planning, the power values reported in Tables 11-3 and 11-4 can be reduced to approximately 90% and 70% of the reported values if the mean number of marked individuals is expected to be close to 18 or 12, respectively. Conversely, the power can be expected to increase if the mean marked number of birds is above 26.3 with about the same coefficient of variation.

References

Fite EC, Turner LW, Cook NJ, Stunkard C. 1988. Guidance document for conducting terrestrial field studies. EPA 540/09-88-109. Washington DC: USEPA.

McDonald LL, Thomas DL, Erickson WP, Krueger HO, Stanley BH, Wetherington JD. 1992. Use of experimental designs in terrestrial field studies: screening studies. Technical Report prepared by Western EcoSystems Technology, Inc., Cheyenne, WY, and Wildlife International, Ltd., Easton, MD.

Miller Jr RG. 1986. Beyond ANOVA: basics of applied statistics. New York: Wiley. 317 p.

Rappole JH, Tipton AR. 1991. New harness design for attachment of radio transmitters to small passerines. *J Field Ornithol* 62:335–337.

Appendix

Notation:

$\bar{c}$	is the average response (e.g., average proportion surviving) over the control sites.

$\bar{T}$	is the average response over the treatment sites.
$s_{\bar{C}}$	is the sample standard error of the controls.
$s_{\bar{T}}$	is the sample standard error of the treatments.
r	is the sample correlation coefficient between the control and treatment sites.
$t_{1-\alpha/2,n-1}$	is the upper $100(1-\alpha/2)$ percentile of the t-distribution with $n-1$ degrees of freedom.
n	is the number of pairs of sites.

A $100(1-\alpha)\%$ 2-sided confidence interval that is exactly consistent with the t-test can be obtained by

$$\frac{X}{Z} \pm \frac{\sqrt{X^2 - (YxZ)}}{Z}$$

where

$$X = \bar{C}x\bar{T},$$

$$Y = \bar{T}^2 - (t_{1-\alpha/2;n-1}xs_{\bar{T}})^2,$$

$$Z = \bar{C}^2 - (t_{1-\alpha/2;n-1}xs_{\bar{C}})^2.$$

Chapter 12

Design principles for radiotelemetry experiments to assess effects on avian survival

John R. Skalski and Steven G. Smith

The essential principles of randomization, replication, and error control found in any well-designed experiment must also be present in field trials in avian toxicology. In addition to these basic requirements, investigators need to take into account the technical demands of conducting radiotelemetry studies to estimate survival rates on free-ranging species. Unique design considerations include choice of experimental unit when exposures can be measured at either the level of the study plot (e.g., spray application) or the individual bird (i.e., blood cholinesterase). Statistical models for assessing survival relationships as a function of treatment conditions and individual exposures will be presented. However, for statistical analyses to be effective, releases and queries of radio-tagged birds must be coordinated between locales and treatments. Examples of commonly occurring flaws in the design and implementation of avian field trials are illustrated along with their consequences on statistical analyses and inferences. Recommendations on the design of multiplot radiotelemetry investigations in avian toxicology to assure sound conclusions are presented.

Among the various uses of radiotelemetry, analysis of survival data has an increasingly important role in risk assessment and the study of selective mechanisms in wild populations. By periodically querying radio-tagged birds, survival rates can be estimated (White 1983) and effects of pesticides and herbicides assessed in a hypothesis-testing framework. Field trials designed to assess the toxicity of pesticides or herbicides on wildlife can be based on replicated control and treatment (i.e., pesticide-applied) plots (Smith and Skalski 1989). Alternatively, individual-based survival models can be used to assess the relationship between exposure, as measured by blood cholinesterase levels, and the fate of individually radio-tagged birds (Hoffmann 1993; Skalski and Smith 1994). Analysis of radiotelemetry data using the program Survival by Proportional Hazards (SURPH) (Smith et al. 1994) can further reveal the interactions between environmental covariates and pesticide exposure to provide a realistic risk assessment. The proportional hazards-based models (Cox 1972) used in SURPH can provide direct estimates of the relative risk of avifauna exposed and not exposed to chemical applications of pesticides and herbicides.

To efficiently assess the effects of agricultural chemicals on wildlife, field trials using radiotelemetry need to be carefully designed, taking into account the statistical models used to estimate survival. Because survival rates are not directly measured, but are estimated across the fates of many tagged birds, coordination of observations among individual and study plots is crucial to a successful field trial. The purpose of this chapter is to discuss key design elements that must be present in the design of field trials using radiotelemetry methods. The consequences of various flaws in study design and potential remedies are discussed. We anticipate that inferential capabilities, precision, and accuracy of survival studies will be greatly enhanced by investigators designing field trials based on the principles presented in this chapter.

Statistical models for survival analysis

Period-specific survival rates can be estimated and associated variances derived for radiotagged birds using the likelihood model (White 1983)

$$L\left(\underset{\sim}{S}\Big|\underset{\sim}{v},\underset{\sim}{a}\right)=\frac{\displaystyle\prod_{i=1}^{K-1}v_i!}{\displaystyle\prod_{i=1}^{K-1}a_{i+1}!(v_i-a_{i+1})!}\prod_{i=1}^{K-1}s^{a_{i+1}}(1-S_i)^{(v_i-a_{i+1})}$$

Equation 12-1

where

K = number of sampling occasions;

a_i = number of previously tagged birds detected alive on occation i ($i = 2,...,K$);

v_i = number of tagged birds known to be alive immediately after occasion i($i = 2,...,K$).

The maximum likelihood estimate for survival in the ith survey period is

$$\hat{S}_i = \frac{a_{i+1}}{v_i}$$

with variance

$$Var\ (\hat{S}_i\ |v_i\ ) \ = \ \frac{S_i\ (1 - S_i\ )}{v_i}$$

Equation 12-1 is suitable for estimating survival rates from a radiotelemetry study at a single population. Equation 12-1 is based on the following assumptions:

1) radio tags remain attached and are not lost by the birds,
2) radio tags do not fail during course of the study,
3) radio-tagged birds remain within the study site,
4) all queries of radio-tagged birds are instantaneous,
5) all tagged animals alive at the beginning of the ith survey period have the same survival probability (S_i) in the period,
6) fate of each tagged bird is independent of all others, and
7) survival probabilities are unaffected by the radio tags.

Smith (1991) used a joint product of likelihoods (Equation 12-1) to investigate effects on survival probabilities at the level of the replicated study plot.

Alternatively, Hoffmann (1993) relaxed assumption 5 to allow each tagged bird to have independent and nonidentically distributed survival probabilities. The resulting likelihood model based on multiple Bernoulli trials for a $K = 3$ period study can be expressed as

$$L\left(S_{ij}\middle|\underset{\sim}{w}, \underset{\sim}{y}, \underset{\sim}{z}\right) = \prod_{j=1}^{N} [1 - S_{1j}]^{1-w_j} [S_{1j}(1 - S_{2j})]^{w_j(1-y_j)}$$

$$\bullet\ [S_{1j}\ S_{2j}\ S_{3j}]^{w_j y_j z_j}$$

$$\bullet\ [S_{1j}\ S_{2j}\ (1 - S_{3j})]^{w_j y_j (1-z_j)}$$

Equation 12-2

where

N	=	number of tagged birds released at beginning of study;
S_{ij}	=	survival probability for the jth bird ($j = 1, \cdots, N$) during the ith period ($i = 1, \cdots, 3$);
w_j	=	$\begin{cases} 1 \text{ if } j\text{th bird survives period 1} \\ 0 \text{ otherwise;} \end{cases}$
y_j	=	$\begin{cases} 1 \text{ if } j\text{th bird survives period 2} \\ 0 \text{ otherwise;} \end{cases}$

$$z_j \quad = \quad \left\{ \begin{array}{l} 1 \text{ if } j\text{th bird survives period 3} \\ 0 \text{ otherwise.} \end{array} \right.$$

In multiplot trials, the effects of pesticide exposure can be statistically evaluated using a joint product of Equation 12-2 across replicate sites.

SURPH program reparameterizes the survival probabilities (i.e., S_i or S_{ij}) using a Cox (1972) proportional hazards model where

"... $$S(t|\underset{\sim}{x}) = S_0(t)^{e^{\underset{\sim}{x'}\underset{\sim}{\beta}}}$$ Equation 12-3

where

$S(t|\underset{\sim}{x})$ =   survival probability from time 0 to t for an individual bird or period described by covariates $\underset{\sim}{x}$,

$\underset{\sim}{S}_0(t)$ =   "baseline" survival probability from time 0 to t when covariate values are zero;

$\underset{\sim}{x}$ =   vector of covariates;

$\underset{\sim}{\beta}$ =   vector of regression parameters for covariates $\underset{\sim}{x}$ ...."

In multiplot experiments (Equation 12-1), the covariates (i.e., x) may describe effects of survey period, plot, and treatment effects as well as environmental covariates. In individual-based models (Equation 12-2), the covariates may describe not only plot and period conditions, but also individual traits and conditions of the birds. Skalski et al. (1993) describes test procedures to determine the significance of plot-wise and individual-based regression coefficients (i.e., β) in tagging models.

Design of a single population investigation

When studying a single population, the objective of the survival study may be 1) to model the estimated period-specific survival rates over time as a function of changing environmental conditions, 2) to model the survival process within survey periods as a function of the traits and exposure levels of individual animals, or 3) to model both extrinsic and intrinsic factors influencing avian survival rates. Regardless of the specific objective, certain minimal design elements are required to effectively analyze the results of radiotelemetry studies and investigate survival processes. Universal among well-designed studies are the following design elements:

1) the study should be conducted during the seasonal period of desired inference,
2) the study should be conducted on the population and at the location of desired inference,
3) the tagged individuals should be a probabilistic sample from the population of inference,

4) each radio-tagged animal should be queried at the beginning of each survey period,

5) querying the radio-tagged animals should be instantaneous, and

6) each survey period should be of equal duration.

Failure to achieve these design elements can result in inferential difficulties at the conclusion of the radiotelemetry study.

The first 3 design elements are common to all well-crafted studies. To obtain valid statistical inferences about a target population, observations must be drawn from that intended population. In field studies, this implies the investigation is conducted at the time and place of the desired inference. Furthermore, the birds selected for radio tagging must be a representative sample of the population intended for inference. Ideally a probabilistic sample of individuals should be drawn using either random or stratified random sampling of the age and sex structure of the population. If a specific subpopulation is to be studied (e.g., fledglings, subadults, adults), then the sample of tagged individuals should be drawn from that stratum. Consideration of the spatial distribution of the individuals is also important. Selecting individuals in only prime or secondary habitat can also bias inferences to the population at large. Failure to properly sample the target population will inevitably jeopardize and bias desired inferences. The degree of bias is related to the degree the sample fails to represent the target population.

The fourth and fifth design elements relate directly to parameter estimation. Failure to query every radio-tagged bird each period can result in biased point estimates of survival and loss of information. Likelihood models (Equations 12-1 and 12-2) assume detection probabilities are 1 (a sure event); failure to query an individual has the potential of confounding the survival process with detection probabilities. For example, a bird queried in periods i and $i + 2$ and found to be alive both periods was certainly alive also in period $i + 1$. In this case the missing datum can be safely ascribed to the omitted query. Alternatively, a bird queried in periods i and $i + 2$ and found to be alive in i but dead in period $i + 2$ could have died during either period $i + 1$ or $i + 2$. Ascribing the death to the wrong period will cause either a positive (i.e., assigning death to $i + 2$ when death is $i + 1$) or negative (i.e., assigning death to $i + 1$ when death in $i + 2$) bias in survival estimates. Currently, the only approach available to data analysis is to censor the data for that individual after period i with the subsequent loss of information. To accommodate known censoring of individuals, likelihood models (Equations 12-1 and 12-2) would have to be modified according to the Kaplan-Meier (Kaplan and Meier 1958) form of analysis. Extending the querying time for the radio-tagged birds to several survey periods has the somewhat similar effect of potentially biasing period-specific survival rates. Observed deaths might be assigned to the wrong period (i.e., period $i + 1$ instead of i) resulting in positively biased survival estimates and less correlation with ambient covariates.

The final design element (i.e., element 6) pertains not to parameter estimation but to interpretation of the survival estimates and relationships between periods. Likelihood models (Equations 12-1 and 12-2) make no assumptions about lengths of survey periods. Valid point estimates of period-specific survival rates are possible regardless of period duration or variation in period durations. However, with varying period durations, length of time is confounded with any changes in environmental covariates under investigation. Consequently, statistical inference concerning correlations between survival and covariates is affected and interpretation of either the presence or absence of significant relationships is impossible. Converting the survival probability (S_i) to instantaneous survival rates; e.g.,

$$S_i = e^{-\mu_i} \qquad \text{Equation 12-4}$$

where

μ = instantaneous natural mortality rate,

t_i = duration of the ith survey period,

does not necessarily mitigate the problem. The reason is that the survival process during a survey period may not be homogeneous, but may be characterized by episodic events in nature. The estimated instantaneous mortality rate μ in Equation 12-4, "averages" the episodic events within a period and potentially diminishes the correlation with causal or correlative covariates.

The analysis of relationships between survival and individual-based covariates contributes additional considerations to the design and analysis of a survival study. Foremost is the nature of the covariates under investigation, whether 1) time invariant or 2) time varying or time dependent. For time-invariant covariates, such as the sex, age class, or nest location, the covariate needs to be measured only once at the time the animal enters the study. Either a single-release or staggered-release design can be used in conjunction with time-invariant covariates. For time-invariant covariates, SURPH analysis estimates the period-specific effects of the covariates.

With time-varying covariates, such as blood cholinesterase activity level, bodyweight, and condition factors, the values change over time and will not stay the same as when the bird was first tagged and measured. Using a single-release design, the regression coefficients for time-varying covariates estimate the diminishing relationship between the independent variables measured at time of tagging and survival in subsequent periods using program SURPH. Using a staggered-entry design, the period-specific releases provide the opportunity to estimate time-specific relationships between the covariates and survival. However, individual releases need to be of sufficient size to provide adequate precision for the estimated survival relationships. With time-varying covariates, the staggered entries act independently and do not necessarily contribute to the overall precision of the study as in the case of time-invariant covariates.

The most common problem we have encountered in the analysis of radiotelemetry data is the failure to query every tagged individual every period. As discussed, depending upon the fate of the missed individuals, biased survival estimates may occur. The next most frequent problem is the loss of animals to subsequent follow up. This can occur because of radio failure or the emigration of an animal from the area. Truncating the data at the period before "loss to follow-up" is necessary to avoid basing survival estimates for the period in which censoring occurred. The occurrence of emigration also puts an alternative interpretation on the estimates of survival, estimating persistence (i.e., survival and continued residence) rather than strictly survival.

An occasional design alternative noted in certain field studies has been the mixing of the querying of birds by radiotelemetry in some periods and use of recapture techniques in other periods. Data from a mixed-mode design cannot be analyzed using a likelihood model (Equation 12-1); rather, specialized likelihood models for release-recapture studies (Cormack 1964) with specific capture probabilities set to the value of 1.0 would have to be used. Mixed-mode models have not been formally developed or published in the scientific literature. The greatest problem with datasets resulting from mixed-mode studies would be the necessity of developing these one-time models tailored to a specific field study. If capture procedures are going to be used to increase sample sizes in the midst of a survival study based on staggered entry, birds already radio tagged should be queried as part of that period's protocol. Both precision and ease of data analysis will benefit from the consistent use of the radiotelemetry capabilities.

Design of multiplot investigations

In multiplot investigations, the individual study plots are the experimental units. The power of statistical tests of survival relationships will depend on the number of replicate plots and number of birds tagged and released per plot. Tests of survival relationships across plots are based on analysis of deviance (ANODEV) procedures (Smith 1991; Skalski et al. 1993). Asymptotic F-tests of survival relationships have the same degrees of freedom as their normal theory counterparts.

Furthermore, the principles of replication and randomization required for valid statistical inference in classical experimentation (Cox 1958; Scheffé 1959; Zar 1984) are also essential for avian field trials. Replication is essential to demonstrate and quantify the extent of the reproductivity of treatment effects. Randomization is essential to assure treatments are not confounded with site effects. Skalski and Robson (1992) discuss inferential requirements for conducting observational, experimental, and impact-assessment studies on wildlife populations. A review of the principles of experimental design is recommended before commencing a large-scale field study.

Several additional key design elements particular to multiplot survival studies are necessary to assure reliable analysis and interpretation of radiotelemetry results. These key design elements specific to multiplot investigations are as follows:

    7) radio-tagged birds must be queried simultaneously in all populations within the investigation, and

8) survey periods must be coincident across all populations within the investigation.

Design elements 7 and 8 are essential in making cross-population comparisons of survival probabilities. Coincident and concurrent survey periods are necessary to assure comparability of survival probabilities between populations.

Failure to retain concurrent and coincident sampling periods is the most common design flaw we have observed in the analysis of multiplot investigations. Omitting survey periods among the various populations results in survey periods of unequal duration between populations. Survival estimates among these periods of unequal duration are not comparable and cannot be statistically compared using program SURPH. Furthermore, treatment effects may be partially or completely confounded with period duration. As mentioned before, converting survival probabilities to instantaneous mortality rates does not eliminate the potential problem of confounding treatments effects with survey period duration, but simply masks the problem.

A less common problem that can occur in multiplot field trials is missing measurements of plot-specific or period-by-plot specific covariates at one or more replicates. The missing independent variables preclude testing the significance of correlations between covariates and survival probabilities for specific sampling periods or the possible exclusion of an entire study plot from the analysis. In the case of time-invariate covariates, a missing covariate measurement can be corrected at a later time with no consequence to the analysis. The situation of missing covariate measurements is much more serious with time-varying covariates. A lost opportunity to collect a time-varying covariate cannot be corrected. The only recourse is to modify the statistical analysis using only a subset of the potential dataset.

Conclusions

Survival studies using radiotelemetry can be efficient approaches to investigating survival relationships at the population and individual levels. However, to make uncluttered inferences concerning survival relationships, traditional principles of experimental design must be followed. Paramount among these design principles are replication and randomization of treatment conditions to study plots. Statistical models for survival analysis dictate additional requirements for parameter estimation associated with the survey technique. The 8 design elements discussed above assure comparability of survival estimates between periods and across populations in order to test relationships with independent variables. Failure to take into account these 8 design elements of radiotelemetry survival studies can result in a loss of information or a confounding of parameter estimates with potential effects of covariates. Ultimately the cost of ignoring one or more of the design elements is the absence of credible statistical inference and the need to repeat the investigation. Attention to design details is crucial for a well-conducted, efficiently analyzed, and clearly interpreted survival investigation. No statistical gymnastics after the study is completed can substitute for a carefully crafted and conducted field study from the onset.

References

Cormack RM. 1964. Estimates of survival from the sighting of marked animals. *Biometrika* 51:429–438.

Cox DR. 1958. Planning of experiments. New York: Wiley. 308 p.

Cox DR. 1972. Regression models and life tables (with discussion). *J Royal Stat Soc B* 34:187–202.

Hoffmann A. 1993. Quantifying selection in wild populations using known fate and mark-recapture designs. Ph.D. Dissertation. Seattle WA: Univ of Washington. 259 p.

Kaplan EL, Meier P. 1958. Nonparametric estimation from incomplete observations. *J Am Stat Assoc* 53:457–481.

Scheffé H. 1959. The analysis of variance. New York: Wiley. 477 p.

Skalski JR, Hoffmann A, Smith SG. 1993. Testing the significance of individual and cohort-level covariates in animal survival studies. In: Lebreton JD, North PM, editors. Advances in life sciences: marked individuals in the study of bird population. Basel, Switzerland: Birkhaeuser Verlag. p 9–28.

Skalski JR, Robson DS. 1992. Techniques for wildlife investigations: design and analysis of capture data. San Diego CA: Academic. 237 p.

Skalski JR, Smith SG. 1994. Risk assessment in avian toxicology using experimental and epidemiology approaches. In: Kendall RJ, Lacher Jr TE, editors. Wildlife toxicology and population modeling: integrated studies of agroecosystems. Boca Raton FL: CRC. p 467–488.

Smith SG. 1991. Assessing hazards in wild populations using auxiliary variables in tag-release models. Ph.D. Dissertation. Seattle WA: Univ of Washington. 351 p.

Smith SG, Skalski JR. 1989. Statistical design and analysis of avian field trials in environmental toxicology. In: Weigmann DL, editor. Proceedings of the National Research Conference on pesticides interrestrial and aquatic environments. Richmond VA: Virginia Water Resources Center. p 407–421.

Smith SG, Skalski JR, Schlechte JW, Hoffmann A, Cassen V. 1994. SURPH.1 Manual.Statistical survival analysis for fish and wildlife tagging studies. Bonneville Cower Administration Division of Fish and Wildlife.PO Box 3621. Portland OR 97283-3621 (http://www.cqs.washington.edu/surph).

White GC. 1983. Numerical estimation of survival rates from band-recovery and biotelemetry data. *J Wildlife Manage* 47:716–728.

Zar JH. 1984. Biostatistical analysis. Englewood Cliffs NJ: Prentice-Hall. 718 p.

Chapter 13

Antenna theory and practices for radiotelemetry applications

Charles J. Amlaner, Jr.

Proper transmitting and receiving antenna selection is fundamentally important to the success-ful application of radiotelemetry in wildlife studies. Transmitting antennas for animal radio-tracking applications in the 30- to 500-megahertz (MHz) range are usually limited to whip-and-loop configurations. In contrast, several types of receiving antennas are routinely used with varying degrees of accuracy. These include the simple dipole, multi-element yagi, H-adcock, omnidirectional whip, and select combinations of the above to form stacked, parallel, collinear, and circularly polarized hybrid antennas. A few automatic-direction-finding receiving systems are in use, and these will be reviewed. Basic antenna theory is discussed, and equations for cal-culating basic dimensional specifications for the more general antennas mentioned above are introduced. Factors relevant to material selection, construction, mounting, antenna coupling, and impedance matching are outlined.

One of the more useful resources written on antenna characteristics and design is the American Radio Relay League (ARRL) Antenna Handbook (1984). This text provides a concise body of information that will enable most students of radio tracking to better appreciate the antenna they depend upon. Amlaner (1980) first summarized the ARRL Antenna Handbook's theory relevant to radiotelemetry applications and also reported on a wide array of antennas specifically used in tracking and biotelemetry. Since that time, several papers have outlined the importance of antenna design and selection in radio tracking (Mech 1983; Kenward 1987; Priede 1992). This chapter expands on the previous contribution and provides an up-to-date review of telemetry antenna design and application.

Radio transmitters produce electromagnetic waves consisting of alternating electric and magnetic fields as part of the transmitters antenna radiation pattern (Jasik 1961). Close to the source of the antenna, there is a near field, and further away there is a far field. Each of these fields has important qualities reflecting its unique application to radiote-lemetry. Mackay (1970) first summarized these qualities (Table 13-1) with respect to field distance, directionality, field strength, and other factors regarding radio-transmitter and receiver-antenna combinations. The near field has a short range and uniform signal pat-tern that starts off strong but diminishes in signal strength very rapidly with distance.

Table 13-1   Contrasting near- and far-field qualities (adapted from Mackay 1970)

	Near field	Far field
Usual use:	Laboratory studies	Field applications
Frequency:	Low (< 30 MHz)	High (> 30 MHz)
Direction:	Omnidirectional	Field has signal nulls
Locating:	Strong vertically or horizontally polarized signal	Maximum signal is perpendicular to field line
Receiver antenna size:	Loop with diameter equal to distance from source	Resonant length dipole, yagi
Field strength:	Decreases by distance^{-3}	Decreases by distance^{-1}

The far field has quite the opposite signal pattern with considerable signal strength beyond several wavelengths at any particular transmitting frequency. It is the far field signal that we measure and that enables us to track animals in the wild. As such, this paper focuses its attention on the practical application of optimizing reception of far-field signals with only occasional reference to near-field applications as they relate to, e.g., effects on antenna performance or laboratory studies.

Basic electromagnetic wave theory begins with the definition of a wavelength expressed by the relationship:

$$\lambda \; = \; v/f \hspace{4cm} \text{Equation 13-1}$$

where (v) equals the velocity of the wave, and (f) equals the frequency of current causing the wave throughout the entire electromagnetic spectrum. For wavelengths in m and frequency in MHz the equation is conveniently expressed as:

$$\lambda_m \; = \; 300/f_{MHz} \hspace{3cm} \text{Equation 13-2}$$

where the velocity of the wave in free space is 3×10^8 m/s. For practical purposes, electromagnetic wave travel through air is almost equal to free-space velocity, thus the differences may be ignored. In contrast, wave travel through water is about 1/9 the speed in air. Equation 13-2 will be referenced throughout this chapter when calculating λ for 30- to 500-MHz antenna element lengths and spacing of these elements along the boom of an antenna.

Several basic measurements may be made on a sinusoidally varying electromagnetic wave, including field intensity (strength), signal polarization, wave attenuation, and signal reflection. Field intensity is measured in terms of microvolts per meter and is the standard measurement for determining the strength of any transmitted radio signal. Rarely is absolute field intensity reported because measurements are tedious, equipment is expensive, and measures must relate to a standard power source. Most radio-tracking applications resort to relative field-intensity measures that can be used in a comparative sense to tell whether or not a change in some aspect of an antenna's design or configuration increased its signal detecting capacity or decreased it (e.g., with receiving antennas).

Antennas composed of several different elements arranged in parallel will have the same polarization as any one of their elements. For example, if all elements on the antenna are vertical, then the antenna will be vertically polarized. Practical experience has demonstrated that vertically polarized signals transmitted from radio-tagged birds tend to penetrate thick forests of primarily vertically polarized trees better than horizontally polarized waves. The horizontal polarized signals tend to travel better under line-of-sight conditions such as over open water or flat terrain.

A radio wave's field strength constantly diminishes as it travels through air. This attenuation is directly related to the distance that a signal must travel, i.e., if a signal has a field strength of 10 μ V m^{-1} at 1 km from the source, then the signal strength will be 5 μ V m^{-1} at 2 km, 1 μ V m^{-1} at 10 km, and so on. There are several factors that will cause a signal to be attenuated much more rapidly than if it were traveling in free space. These include the density of the medium (e.g., air is approximately 1/9 the density of water) and in a very practical sense, the line-of-sight path the signal takes between transmitting and receiving antennas. The estimated transmission distance in kilometers to the horizon (D_{km}) due to the curvature of the earth is given by the following equation:

$$D_{km} = 4.124 \, (H_m)^{0.5} \qquad \text{Equation 13-3}$$

where H_m is the antenna elevation in m above ground. This equation takes into account that under normal conditions a slight bending of the radio signal occurs over the earth's sphere. Generally, antennas should be placed as high as possible above ground and should not be near or adjacent to steep, irregular terrain, power lines, wire fences, buildings (especially metal structures), or thick forests. Power lines, wire fences, and thick forests tend to bend radio signals, which seriously affects signal transmission distance and directional accuracy of the receiving antenna. If an antenna is placed on a hill, it should not be located at the top, especially if the transmission originates in front of a hill with a steep slope. Radio waves arriving at such hills tend to be reflected above the antenna if it is placed on the summit. This problem can be alleviated to a great extent by elevating the antenna higher above the ground surface.

Another form of signal attenuation is attributable to the human body when it is in very close proximity to the receiving antenna (Neukomm 1978; Neukomm et al. 1981; Klaus et al. 1982). While animal tissue attenuation usually is not considered an important factor in antenna performance (except in collar-loop antennas), it is reported to be a significant attenuating factor in near-field performance and a lesser factor in the far field (Neukomm 1978; Neukomm et al. 1981). It appears that the body affects the near field by giving antennas closely spaced to the body a more uniform, omnidirectional pattern if they are operating near body resonance (i.e., ½ λ resonance is approximately 75 MHz). Neukomm (1978) reports that antennas coupled closely to the body will be relatively uniform in this effect within the 50 to 150 MHz band. If a human body (a conductor with several ionic salts in solution) comes between the transmitter source and antenna, there will be a 1 to 4 dB decrease in signal strength. It is now important to keep in mind the effect our bodies have on the relative signal pattern of an antenna and how this will diminish the antenna's directional accuracy in actual field usage.

Reflection of electromagnetic waves is well known. Just as you view your image in a mirror, radio waves can reflect on certain surfaces to a greater or lesser degree. This phenomenon is often referred to as a "bounce." Its presence can create havoc for accurate radio direction determination because it is a prime cause of multipath, the mixing of many reflected signals. Reflected radio signals are commonly found in mountainous regions and also, to a lesser extent, in wet forests. The unfortunate result is that an animal is thought to be in a particular direction which is quite the opposite of practical knowledge of "where the animal should be located." Garrott et al. (1986) evaluated the effects of reflected signal bias in triangulation. The only way of eliminating bounce as a possible confounder is to move to another location and take additional azimuths.

Consider now the association between transmission and receiving antennas. For practical purposes, the properties of a tuned receiving antenna are the same as those used in transmission. The gain, pattern, and impedance of both receiving and transmitting antennas are the same at any point of measurement. The only conceptual difference is that a receiving antenna acts to intercept the usually minute source of radio power, whereas in transmission it is the actual radiating structure for the transmitter.

Transmitting antennas

Whip, core, and loop antennas are commonly used transmitter antenna configurations. The efficiency of a simple loop antenna is greatly increased by concentrating the electromagnetic flux, and this can be achieved by using a ferrite core within the loop itself (a configuration commonly called a "dustcore"). Various sizes of ferrite cores are available commercially with approximately a 2- to 4-mm diameter being standard. A gain factor of approximately 10 can be achieved by using a ferrite core instead of a simple loop of the same diameter. The electromagnetic wave pattern from loop and dustcore antennas approximates a "figure-8 donut" or toroid that gives it a definite signal polarization. Reduced size is the main advantage of using loop and dustcore antennas, which makes them particularly useful in implanted transmitters (Amlaner 1978). Larger-diameter loop antennas made from wire braiding may be incorporated into a transmitter's collar or harness. However, this configuration must be tuned to resonance to optimize power output by placing an appropriate value of capacitance (e.g., a variable capacitor) in parallel to the collar's loop antenna. Before deploying this antenna configuration, it is a common practice to tune the loop antenna on an animal model that reliably simulates size, mass, and tissue composition. This pre-tuning helps compensate for stray capacitance created by the interaction of the radio-tagged animal and antenna, which otherwise might cause detuning and poor transmitter performance.

Straight wire antennas (whips) are more efficient signal radiators than loop and dustcore antennas. The selection characteristics of the whip antenna are affected by several significant factors including 1) behavioral and physical characteristics of the animal, 2) transmission frequency, 3) method of mounting the transmitter on the animal, and 4) method of mounting the antenna on the transmitter. Whip antennas are particularly suited for bird telemetry, since there are proven methods for attachment (Anderka and Angehrn 1992). Some of the common configurations include tail mounting the radio pack, thereby incorporating the antenna and transmitter into the tail feathers (Kenward 1987) and backpack mounting using a harness made from cloth covered elastic band (Amlaner et al. 1978). In the backpack configuration, the antenna can be either horizontally (more common) or vertically mounted (when transmitting at UHF the antenna is very short). Whip antennas are usually constructed with spring stainless steel wire. Solid or stranded types may be used depending on the degree of antenna movement and the animal's ability to destroy the antenna from preening and flight. Fine gauge piano- or guitar-string wire are often used, but a marine grade stainless steel cable covered with Teflon or PVC insulation is preferred. The calculation of l for a whip antenna is given in Equation 13-2. Most practical applications will necessitate that the antenna length be as short as possible; therefore, optimal wavelengths of ½ or ⅝ λ cannot be achieved and tradeoffs must be found.

Power output is greatly dependent upon the efficient coupling of a radio frequency (RF) section to the transmitting antenna. The best antenna-RF coupling is only achieved when the impedance of the antenna closely matches the RF output impedance of the transmitter. From Ohms Law we know that antenna impedance (measured in ohms, Ω) is equal to the voltage applied at the antenna divided by the current flowing into the antenna. When the antenna is resonant, the current and voltage will be in phase and the impedance will be purely resistive. This situation normally produces the strongest field strength (output power) from an antenna. Through experiments, we can determine the best antenna impedance matching, but it will usually require some method of measuring relative signal strength with an RF power meter. If the transmitter RF stage has a facility for tuning the antenna to resonance, which in practice optimizes the power output, then the longest practical length of antenna should be used. This is usually impossible to achieve in ultra-miniature transmitters that utilize fixed values of capacitance in the RF section. On these transmitters, a ¼, ⅛, or 1/16 λ antenna length is usually chosen and subsequently tuned by clipping the antenna wire; however, the antenna's efficiency is often dramatically reduced at these shorter lengths. For example, at 1/16 λ, transmission range for a low power RF transmitter may be only a few 100 m. Whip antennas may be shortened without a significant impedance mismatch by the use of a loading coil between the RF stage and the base of the antenna. Loading coils are made from 2 to 6 turns of enamel-coated wire (20 to 40 gauge) that may be encapsulated within the transmitter.

Receiving antennas

Differing laboratory and field applications require specific receiving antennas. For example, long-range reception may be desirable but the large antenna arrays usually required for high gain applications cannot be carried very easily; therefore, a compromise must be made between gain and portability of the antenna system. Project aims and telemetry operating frequency (e.g., tracking fish using high frequency, terrestrial radio tracking of birds and mammals using VHF and UHF), will largely dictate the receiving antenna's shape and size or the following sections describe the common receiving antenna configurations used in radiotelemetry, along with some suggestions on their construction and mounting.

Directional antennas

The antennas discussed in this section all have a plane in which the signal strength is greatest (lobes) and a secondary plane in which the signal strength is minimal (nulls). Exploiting these antenna characteristics provides useful information about direction and signal strength of a transmitter (for practical applications, see Lancia and Dodge 1977; Amlaner 1978; Macdonald and Amlaner 1980; Mech 1983; White and Garrott 1990; Gonczi 1992).

Dipole antennas

The ½ wavelength dipole is the basic receiving antenna configuration used in radio tracking, and its signal gain is the standard to which all other antenna configurations (e.g., yagi, H-adcock, etc.) are compared. The dipole antenna is small and readily portable at VHF frequencies and higher which makes it appropriate for telemetry field applications. Dipole antenna designs have been published (Burchard 1988a, Kenward 1987) with some versions being foldable or collapsible (Parish 1980; Bosak 1992). Most of the telemetry manufacturers produce dipoles which are custom tuned for the specific frequencies of their transmitters. A dipole's minimum signal strength is directly in line with the axis of its 2 elements, and the strongest signal is perpendicular to this axis, thereby producing a 3-dimensional figure-8 donut or toroidal pattern. This symmetrical antenna pattern is omnidirectional and offers no directionality with which to determine the direction of signal arrival. However, often antennas are not perfectly aligned or tuned, and the user may detect subtle differences between the front and back of the dipole and can usually achieve tracking azimuth accuracy between 10 and 20°. The dipole resonant length (½ λ) can be calculated from

$$\tfrac{1}{2}\,\lambda_m = 150 / f_{MHz} \times k \qquad\qquad \text{Equation 13-4}$$

where k is a constant, calculated from the ratio

$$k = \tfrac{1}{2}\,\lambda_m / d_m \qquad\qquad \text{Equation 13-5.}$$

Here d equals the diameter of material used to make the dipole. A ratio of 10,000 yields a k value of approximately 0.98 that effectively means smaller ratios yield shorter electrical lengths of the antenna (Amlaner 1980, page 254) (Figure 13-1). In practical applications, metal tubing and solid rod of similar material and dimension are considered to have the same constant k.

Loop antennas

These antennas are useful in applications where the tagged animal-to-antenna distance is short and the antenna is fixed. Loop antennas are often used in freshwater fish-tracking applications where radio frequencies are 30 to 75 MHz (Priede 1992). The resonant circumference of the loop, for frequencies near 30 MHz and with k large (see above), is calculated as

$$C_m = 306.3 / f_{MHz} \qquad \text{Equation 13-6.}$$

A signal may be received from almost anywhere within the circumference of a horizontally polarized loop. A vertically polarized loop has directional sensitivity when the transmitter is outside the loop's perimeter. Maximum signal strength is in line with the plane of the loop, and the pattern has symmetrical front and back lobes shaped like a figure 8 donut similar to a dipole's pattern. The minimum signal (null) occurs when you look through the loop directly at the signal source. Our experience has shown that vertically polarized directional azimuth accuracy is approximately 5 to 15°, slightly better than the dipole; however when the loop is horizontally polarized, directionality is severely reduced. The theoretical signal gain in comparison to a half-wave dipole is approximately +3 dB. Loop antennas can be designed for higher frequencies with the addition of a tuning capacitor (see transmitter-tuned loops above). Tuned loop antennas are often adapted to laboratory applications (Vreeland et al. 1976). Three loops placed at orthogonal axes that are sequentially selected by a receiver give reasonably complete signal coverage of an animal freely moving within a room or building (Mackay 1970).

Multielement antennas

Radio-tracking field studies probably use more multielement receiving antennas than any other antenna configuration (Dunstan 1977; Balwanz 1978; Amlaner and Macdonald 1980; Burchard 1987, 1988b; Gaudin et al. 1991; Morris 1992). These antennas offer the advantage that, at higher frequencies, added elements increase the antennas' gain. Also, directional antennas tend to concentrate reception of very weak signals over a much narrower angle (lobe) than other antenna designs. Generally, azimuth directionality is improved to within 0.5 to 5° when transmitters are in far-field distances of greater than 20 λ. Radio-tracking applications of multielement antennas are usually only suitable for frequencies well above 30 MHz due to significant element sizes and boom lengths required for operation at low frequencies.

Optimal multielement design will take into account the antenna's overall gain and front-to-back ratio (a measure of gain difference between front signal reception by the antenna and back signal reception). To obtain a narrow capture area, and hence good front lobe directionality, the front-to-back ratio should be optimized so that back signal reception is several dB less than front signal gain. In practice, there are various compromises between optimizing antenna gain over front-to-back ratios. These are determined by adjustments in element spacing, number, and length of elements; these factors are dictated to a large extent by the specific research application.

The different antenna elements as outlined in Figure 13-1 are 1) the driver, which is the element directly wired to a transmitter (when sending power) or the receiver (when receiving power); 2) the parasitic elements, which direct power through electrical coupling to the driver; and 3) the reflector, which couples energy to the driven element and is located at the back of the antenna. The addition of a reflector to a simple dipole (driven element) causes the front lobe to narrow and have better signal capture (higher gain) than the back lobe. The addition of parasite elements causes further narrowing of the front lobe and greater increases in gain. The H-adcock (Taylor and Lloyd 1978) and yagi (Uda and Mushiake 1954) are special types of multielement antennas.

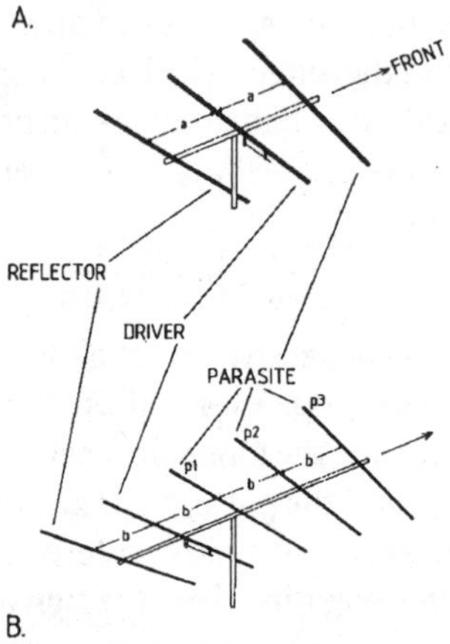

Figure 13-1a and 13-1b
Figure a Three-element directional yagi on 102 MHz has an element spacing (a) = 43.7 cm, reflector length = 145.5 cm, driver = 137.4 cm, and parasite = 130.6 cm.
Figure b Five-element yagi on 173 MHz has an element spacing (b) = 35.5 cm, reflector length = 86.4 cm, driver = 81.3 cm, p1 = 76.2 cm, p2 = 75.6 cm, p3 = 74.9 cm.

H-adcock or "Lazy H" antenna

The H-adcock antenna (Figure 13-2) is commonly used in radio tracking because it has good directional azimuth qualities for a portable antenna (2 to 5°) with moderately high gain (4.4 to 6.7 dB) over the half-wave dipole, and it may be designed to fold or collapse when not in use. Table 13-2 shows the benefits of various fractional spacing of the 2 elements relative to antenna gain. In practice, ½-λ-element spacing is optimum for frequencies in the 100 to 150 MHz range. When the elements of the H-adcock are vertically polarized, 2 symmetrical lobes will be directed maximally in line with the boom, and nulls will be perpendicular to the boom. Because all elements on the H-adcock are symmetrical, there is no difference between the front and back lobes, much like the dipole and loop antennas; thus it is again difficult to distinguish direction without field experience. The H-adcock antenna must be connected to a receiver through an impedance matching circuit (balun match) in contrast to the dipole, which if designed properly may be connected directly to a receiver with standard 50Ω coax cable. It is critical that each balun loop length be precisely measured to maintain the antennas directional accuracy and impedance matching. A cable velocity factor (see Equation 13-11) must also be taken into consideration when computing these loop lengths.

Yagi or Yagi-Uda antenna

A multielement array with linear dipole elements was earlier described by Uda and Mushiake (1954) and is generally known today as the "yagi" antenna. This antenna has extremely good directionality and has the highest gain of any antenna discussed so far, especially when several parasitic elements are employed in its design. Yagis are the most widely used antenna in radio tracking research because of their simplicity and high gain (Amlaner and Macdonald 1980; Kenward 1987; Burchard 1988a). The 2-element yagi requires only a driver and a reflector (Burchard 1988a). Several commercial telemetry suppliers provide fold-up versions of 2- and 3-element yagis, which are the antennas of choice for backpacking and long-distance radio tracking on foot. Overall, these antennas provide the best balance of signal gain and front-to-back ratios, and they are light weight. Stationary (fixed) antenna systems routinely employ from 3- to as many as 20-element yagis, which are horizontally or vertically polarized. We found that a single 7-element yagi, horizontally polarized, provided excellent gain and directionality for long-range bird tracking where line-of-sight conditions usually prevailed. Several yagi antennas can be mounted (i.e., stacked) side by side to obtain additional signal gain, better directionality (compared to a single antenna), and the possibility of null-peak reception when properly fed by matching networks to a receiver. Practical experience has shown us that 2 5-element yagis stacked side by side, separated by approximately 1 λ and vertically polarized, provided excellent gain and directionality for vehicles or boats. These antenna systems achieve excellent azimuth directionality when properly tuned (± 0.5°) and fed through a null-peak matching network (Figure 13-3).

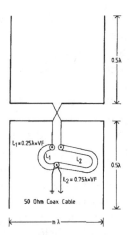

Figure 13-2 An H-adcock or Lazy H antenna with balun impedance matching network. The dimensions are given in Table 13-2.

Table 13-2 Gain of an H-adcock antenna compared with a dipole under different element spacing fractions of λ. See also the H-adcock antenna design figure.

3/8 – wave spacing	+ 4.4 dB
1/2 – wave spacing	+ 5.9 dB
5/8 – wave spacing	+ 6.7 dB
3/4 – wave spacing	+ 6.6 dB

Yagi construction

Effective yagi design takes into account 1) element length, 2) spacing between elements, and 3) the total number of antenna elements. Element lengths of multielement antennas are approximately ½ λ ± 5%, and spacing between elements may vary from 0.1 to 0.2 λ.

Figure 13-1b shows a typical 5-element yagi antenna with a reflector, a driven element, and 3 parasites. Theoretically, there is no limitation to the number of parasites that may be used. At VHF and UHF ranges, multielement antenna systems exceeding 10 elements are common; however, the practical benefits of increased gain versus overall boom length diminish beyond 20 parasite elements. Figure 13-4 shows the relationship between the number of antenna elements and antenna gain in dB above the gain of a half-wave dipole.

Published element lengths and, to a less critical extent, element spacing may not conform to the values calculated to give maximum gain. The practical compromises are usually in terms of sacrificing gain for a greater antenna-receiving band width (i.e., frequencies received above and below the designed center frequency) or increasing the front-to-back ratio to produce more signal gain. To achieve these specifications, parasitic elements should be approximately 5% less in length than the driven element, progressively decreasing in size by 5% for each parasite mounted on the array. The reflector element should be 5% longer than the driven element. Element diameter should be no more than 1.5 cm (except the driver element); and the boom, 2.5 to 4 cm. For higher frequencies (above 200 MHz), the elements may be made from solid rod (< 0.5 cm). The driven element should be close to 2 8 in length and 2 to 3 X the diameter of the other elements. However, commercially available, foldable yagi antennas usually use uniform diameter tubing for all elements.

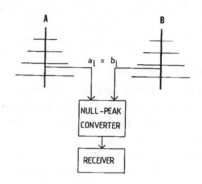

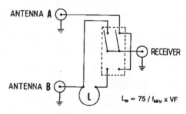

Figure 13-3   *The block diagram and schematic of a null-peak dual yagi configuration. The 2 coax cables a_l and b_l must be identical in electrical length. Dashed lines on the schematic represent a low-loss double pole, double throw switch. VF is the velocity factor of the coax used in the loop (L).*

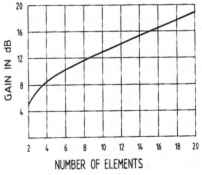

Figure 13-4   *Gain of a yagi antenna compared to a half-wave dipole versus number of elements on the yagi. Redrawn with permission of the ARRL.*

These antennas can conveniently be made from aluminum tubing found at most electrical or plumbing suppliers. For permanent installations or in corrosive marine environments, marine grade stainless steel tubing is preferable to aluminum.

When designing your own custom yagi antenna, the lengths of each element type (where gain is maximized and receiving bandwidth minimized) may be calculated by using the following equations:

$$\text{driver element}_m \quad = \quad 146.0 \, / f_{MHz} \qquad\qquad \text{Equation 13-7}$$

$$\text{reflector element}_m \quad = \quad 152.4 \, / f_{MHz} \qquad\qquad \text{Equation 13-8}$$

$$\text{first parasite}_m \quad = \quad 138.7 \, / f_{MHz} \qquad\qquad \text{Equation 13-9}$$

$$\text{second parasite}_m \quad = \quad 138.7 \times 0.95 \, / f_{MHz} \qquad\qquad \text{Equation 13-10}$$

and so on for additional parasites. Element length calculations take into account the *k* factor discussed above, which essentially decreases overall physical length of each element. Equations 13-7 through 13-10 use an average length/diameter ratio of 200 to 400 (*k* = 0.96 to 0.97) and element spacing of 0.1 to 0.2 λ.

Multielement signal strength patterns and tuning

The practical requirement of all telemetry antenna configurations is to reliably receive or transmit a signal over a well-defined pattern. There is no distinction between radiation and reception patterns for antennas, but the reception patterns that are significant for telemetry receivers will be discussed. Figure 13-5 summarizes the horizontal plane patterns for each type of multielement antenna in comparison to a dipole. Each reception pattern is based on several assumptions: 1) the antenna is well above ground level (10 to 20 λ) to minimize effects caused by ground conduction, 2) the array is properly tuned, 3) a signal source is more than 20 λ away, and 4) each element on the array is straight and is the proper dimensions for its frequency bandwidth (broken and bent elements cause significant alterations to a reception pattern). On bidirectional arrays, the major and minor signal lobes are symmetrical (e.g., H-adcock). On multielement antennas of 2 elements or more (e.g., Yagi-Uda or variation), the unidirectional array will have a prominent major lobe (the front) with several minor lobes radiating to the sides and behind the major lobe. Notice that null-peak dual yagi systems have a prominent center null. We have found that a null is easier to detect when the signal is very distant and weak rather than attempting to discriminate the peak of the lobe pattern (see Macdonald and Amlaner 1980 and Kenward 1987 for details on radio tracking with yagi systems).

The yagi antenna (as well as other antenna configurations) should be tuned for a minimum voltage standing wave ratio (VSWR) by tuning the trombone slide for maximum signal (see gamma match below). This is achieved by using either a VSWR bridge (an inexpensive version may be obtained from Radio Shack Electronics or other electronics stores serving the ham radio industry) or by tuning the gamma match empirically while listening to a continuous wave transmitter located greater than 20 λ's away from the antenna. It is important to note that any conductive material placed near an antenna's elements anywhere on the array may cause severe detuning effects that result in lower antenna gain and directional accuracy. Antenna elements should not be closer than ½ λ to the nearest conductor (e.g., the roof of an automobile). Holding or mounting an antenna by its elements also causes detuning.

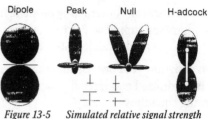

Figure 13-5   *Simulated relative signal strength patterns of the common yagi configurations compared to a dipole and H-adcock. These patterns represent measurements under ideal conditions as outlined in the text. The signal*

Omnidirectional antennas

Where animals are moving within a confined area, loss of a transmitter's signal is minimized by using an omnidirectional antenna that is capable of receiving signals in a 360° pattern with relatively uniform gain. A simple omnidirectional antenna consists of 4 equally spaced elements acting as an artificial ground plane evenly spaced around the ¼ λ radiator (Figure 13-6). This antenna may be end fed directly by 50 W coax cable with only a slight impedance mismatch. These antennas are exceptionally useful in radio-tracking projects. For example, if an animal's radio signal has been lost, a ¼ or ½ λ whip omnidirectional antenna may be attached to the roof of a vehicle (which acts as the ground plane) to facilitate locating the signal.

Once the general area is found, a yagi can be remounted on the vehicle for traditional azimuth radio tracking. We routinely use omnidirectional antennas attached to remote monitoring stations for receiving signals from animals in the wild. In this application, we note when the animal is present or absent (e.g., bird nest attendance, sleep sites, etc.) from the receiving field of the antenna. An appropriate recording device (e.g., strip chart recorder, tape recorder, or portable computer) logs the data that are collected at a later date.

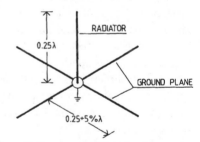

Figure 13-6   *An omnidirectional ground plane antenna with approximately 50Ω impedance.*

Matching the antenna to coaxial cable

Techniques used for matching the coaxial cable impedance and minimizing VSWR on receiving antennas include the gamma match, the T match (a special type of gamma match), and the balun match. The reason for matching coax cable to the antenna is that at various frequencies the antenna exhibits a different impedance (see above) which the cable sees as it receives the signal. There is usually a compromise between matching impedance (thus minimizing VSWR) and allowing the antenna to operate over a wider frequency bandwidth. If we minimize the VSWR, this tends to maximize gain but to minimize the band width of the antenna as well. If the VSWR is not minimized, then gain by the antenna is greatly reduced with a corresponding change in the radiation pattern. Matching impedance is achieved by characteristics of the matching network: length and spacing for the gamma match (Figure 13-7) or length of the loops for the balun (Figure 13-2).

Gamma match construction

The gamma match (Figure 13-7), or trombone slide, is used routinely on yagi antennas to match the impedance of feed cable (e.g., 50 and 75 Ω coax) to the impedance characteristics of the antenna. The gamma match for each yagi in Figure 13-1 can be made from the same material as the elements. The gamma match should have a capacitor (50 pf) placed in series with the feed line for 50 MHz and the metal tubing length

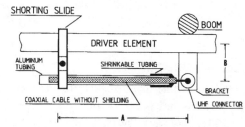

Figure 13-7   *The gamma match or trombone slide used on yagis of Figure 13-1 to match 50 or 75 Ω coax cable. See text for explanations about dimensions A and B.*

should be 30 cm. At 150 MHz, the gamma match should be 13 cm long and should use a 14-pf series capacitor. Antennas using these measurements can be fed directly with 75 Ω coax cable. However, for matching a multielement array to the common 50 Ω coax cable, the length of the gamma match rod should be 0.04 to 0.05 λ (Figure 13-7, dimension A), an outside diameter ⅓ that of the driver diameter, and center to center spacing with the driven element approximately 0.007 λ (Figure 13-7, dimension B).

Coaxial cables

Common types of receiving cable are RG58U, RG59U, RG11, and RG8. These coaxial cables have an impedance of approximately 50 Ω (RG58 and RG8) or 75 Ω (RG11 and RG59). At frequencies above 30 MHz and up to 200 MHz, the attenuation through 30 m of cable is approximately 2 to 3 dB. On this basis alone, the length of cable between receiver and antenna should always be kept to a minimum so as not to nullify the advantages of a high gain antenna. Antenna cables should be frequently checked for abrasion and wear due to rough field use and the effects of weather on the insulation. Frayed or worn cable should always be replaced as increased signal attenuation occurs when the insulation breaks down. The same recommendation is given for broken or weak bayonet coupling (BNC) or UHF connectors, which are the common terminations made on coax cable. It is almost impossible to replace a broken connector in the field and, as such, it is wise to always carry a spare coax cable with proper connectors for the antenna and receiver.

Electrical length of coax cables

Whenever reference is made to a coax cable being a multiple of 1, as calculated by Equation 13-2, we refer to its electrical length, not its physical length. Electrical cable length is given by the equation

$$\text{electrical length}_m = 300 \,/\, f_{MHz} \times \text{cable velocity factor} \qquad \text{Equation 13-11.}$$

This velocity factor is a ratio of the actual signal velocity along the coax to the signal velocity in free space. The velocity factor for the 4 common types of coaxial cable mentioned above is approximately 0.66. For other cable velocity factors, consult the manufacturers. Note that a velocity correction was made for the balun matching loop in Figure 13-2.

Antenna mounting

Smaller antennas may be held by hand as long as the effect of body conductance does not severely interfere with reception. All antennas discussed in this chapter may be mounted on poles, booms, or masts, using U bolts or stainless steel hose clamps. Again, it is important to maintain a reasonable distance (> ½ λ) between the antenna array and supporting structure if it is a conductor. We recommended that arrays be mounted using reinforced PVC pipe (a non- conductor) for the first 1 to 2 λ; then secure the PVC to aluminum pipe or other light weight structure. For applications requiring 3 m height or more, inexpensive inter-locking aluminum mast sections in 3.5- and 7-m lengths can be purchased from most radio and television repair shops. Antenna rotators can be mounted on top of the mast to facilitate slewing the array for taking azimuths while radio tracking (e.g., Smith and Trevor-Deutsch 1980). Mounting antennas on vehicles, boats, and aircraft requires secure attachment of the array and supporting structures. Vehicle and boat installations often use guy wires supporting the mast and attached to a base (Kenward 1987) or a portable base with welded supports may be installed in the bed of a truck (Amlaner 1980). Aircraft mounts for 2- and 3-element yagi systems require extra secure mounting plates, which must be Federal Aviation Administration (FAA) approved (for government research and to maintain aircraft insurance). Kolz (1991) recently secured FAA certification for a simple wrap-around mount for the Cessna 172 and is currently seeking approval for the Cessna 150, 2 aircraft commonly used for radio tracking. Mech (1983) and Kenward (1987) give examples of noncertified mounts as well as excellent recommendations on how to track from aircraft using null-peak yagi arrays.

Automatic antenna selectors and direction detection

For antenna systems employing several arrays without mechanical antenna rotators, the problem remains how to optimally select an individual antenna for determining an azimuth without human intervention. Two approaches have been reported: 1) rapidly selecting one of several antennas through an electronic switching network that also detects the strongest signal-antenna combination (e.g., Mackay 1970; Groeneveld and de Bakker 1982; Kuechle et al. 1983 [aircraft system]; Lemnell et al. 1983; Yasuda and Inai 1990) and 2) rapidly sampling an array of vertically polarized dipoles while detecting a Doppler shift or phase relationship of the received signal, which is translated into a signal azimuth (Amlaner 1984; Balendonck et al. 1990; Angerbjorn and Becker 1992). In most cases, 3 or more automatic systems are required for reliable automatic azimuth triangulation (White and Garrott 1990). A few universities and government research centers have installed multiple fixed antenna systems that incorporate hybrid versions of the technology mentioned above (e.g., Grimso, Sweden; Chize, France; Ceder Creek, USA; Wageningen, Netherlands). These systems provide round-the-clock monitoring of animal locations and, in some cases, physiological monitoring.

Practical radio-tracking exercises and antenna accuracy

Because we rarely see our radio-tagged animals, we must have considerable confidence and understanding about the accuracy of the antenna that we rely on for azimuths. One of the best ways to build confidence in your antenna system is to practice frequently with transmitters placed in known locations and track them using standard triangulation techniques. A blind radio-tracking exercise is best, where another researcher places the transmitters without knowledge by the tracker. By placing transmitters in locations which create bounce, multipath, signal polarization, etc., the tracker learns to cope with real difficulties while gaining confidence in the antenna's directional sensitivity. Once a tracker can reliably obtain azimuths within acceptable experimental errors, usable data may be collected. From time-to-time thereafter, refresher exercises should be conducted to verify further the precision of the tracker and the antenna.

For each antenna type mentioned in this chapter, the range of potential azimuth accuracy is based on actual field experience. Radio tracking is not a precise science; rather, it is more of an art that is highly dependent upon the ability of the tracker and quality of the equipment used. Under ideal field conditions, antenna accuracy will be close to theoretical values, but ideal field conditions are experienced rarely. To better appreciate the problem of assessing field accuracy, readers should reference Macdonald and Amlaner (1980), Hupp and Ratti (1983), Kenward (1987 Chapter 6), and White and Garrott (1990 Chapters 4 and 5).

There are 3 categories of error that can frustrate the tracker (Macdonald and Amlaner 1980): 1) system error, inherent in the receiving equipment and its operations as outlined above, 2) movement error, resulting from the animal's moving while azimuths are taken, and 3) topographical error, resulting from anomalies, such as reflection and refraction of the radio signal creating false azimuths, as outlined above. System errors are caused by 1) inaccuracies in the directionality of the antenna as a result of broken or bent elements, antenna leads, and connectors; 2) inaccuracies imposed by the geometry of triangulation while taking an azimuth; and 3) inaccuracies resulting from misreading the azimuth, e.g., because of holding a magnetic compass too close to a vehicle or antenna. Table 13-3 shows the magnitude of system error on the accuracy of azimuths taken with a dipole and H-adcock on the position of a sleeping fox lying on the edge of woodland and open habitats. Note that system error does not increase significantly with distance from the fox and is 3 to 5°. The error is significantly greater for azimuths taken from within the woodland to those taken from open farmland. The dipole's average error was marginally worse than the H-adcock's. A similar test under urban conditions yielded far less accurate azimuths, a result attributed to considerable multipath and other building interferences. While the system error does not increase with distance, the triangulation error does. The area encompassed by the beam of inaccuracy increases with distance from the animal. With a 3° error, this constitutes a decrease in the accuracy of the supposed position of 52 m km^{-1} calculated by

Table 13-3   *A comparison of the mean angular error of sets of azimuths taken from different positions with a dipole and H-adcock antenna under ideal conditions (near = 100 m and far = 500 m approximately)*

		Azimuth errors (degrees)		
		Woodland	Field	Town
H-adcock	Near	4.2	3.1	20.9
	Far	5.1	3.3	20.1
Dipole	Near	7.4	3.8	12.0
	Far	12.6	3.9	25.1

$$\tan \theta \times \text{distance} = \text{displacement} \qquad \text{Equation 13-12.}$$

Furthermore, error polygons increase in area for a given distance between transmitter and receiver, depending on the extent to which the intersecting azimuths depart from right angles (and hence the intersecting beams of inaccuracy become less of a square and more of a parallelogram). Thus, azimuths should be taken at approximately right angles to each other and as close to the animal as possible.

Movement errors depend upon a delay between taking an azimuth and the animal's movement. These errors can be minimized by rapidly taking azimuths, so as to minimize the effect of animal movement or to employ multiple tracking sites for animals that move rapidly.

Topographical errors are the most serious. Although not commonly stressed in the literature, the radio tracker's subjective feel for the radio landscape of his study area is widely acknowledged as an important facet of accurate radio tracking. The following suggestions will help to minimize the risks of both movement and topographical errors (Macdonald and Amlaner 1980):

1) Take consecutive azimuths with as little delay as possible. Often in direction finding, the nulls are easier to detect; and if the antenna has a pattern of sharp nulls, these can be used to quickly determine the azimuth.

2) Know the individual features of the landscape and how they affect radio signals.

3) Take several azimuths if errors are suspected; at least 3 are necessary to detect movement errors.

4) Rank signal quality, azimuths, and the derived radio locations in terms of their accuracy.

Keeping alert for erroneous locations is difficult in the field unless the tracker is aware of these potential problems and their potentially serious results.

Telemetry receiver-antenna-transmitter system summary

It should be obvious that a telemetry system's weak link exists when the transmitter antenna, feed line, or receiver are not properly matched and tuned. It is beyond the scope of this chapter to outline all of the many parameters that are necessary to specify a quality radiotelemetry system. It is equally difficult to specify the many ways a system might misbehave. Kuechle et al. (1981) reviewed the specification standards and measurement techniques for evaluating radio-tracking receivers. Kuechle et al. masterfully outlined how the combined effects of atmospheric noise, antennas, cables, transmitters, receivers, and operators each contribute to the efficiency of a telemetry system. With increasing levels of government involvement in evaluating telemetry system performance, it is also important that all parts of the system perform as expected; Beach and Storeton-West (1982) outlined their experience while obtaining government approval for a radio-tracking system.

Acknowledgments - I want to express sincere appreciation to my graduate students, post docs, and colleagues—especially A. L. Kolz—for sharing their rich radio-tracking experiences with me over the last 20 years. The American Radio Relay League gave kind permission to use Figure 13-4.

References

American Radio Relay League Antenna Handbook. 1984. The ARRL antenna book. 14th ed. Newingon CT: American Radio Relay League.

Amlaner Jr CJ. 1978. Biotelemetry from free-ranging animals. In: Stonehouse B, editor. Animal marking: recognition marking of animals in research. New York: Macmillan. p 205–228.

Amlaner Jr CJ. 1980. The design of antennas for use in radio telemetry. In: Amlaner Jr CJ, Macdonald DW, editors. A handbook on biotelemetry and radio tracking. Oxford: Pergamon Press. p 251–261.

Amlaner Jr CJ. 1984. Taking laboratory experiments into the field: studies of sleep in wild animals. In: Kimmich HP, Klewe HJ, editors. Biotelemetry VIII.

Amlaner Jr CJ, Macdonald DW, editors. 1980. A handbook on biotelemetry and radio tracking. Oxford: Pergamon. 804 p.

Amlaner Jr CJ, Sibly R, McCleery R. 1978. Effects of transmitter weight on breeding success in herring gulls. *Biotelem Patient Monit* 5:154–163.

Anderka FW, Angehrn P. 1992. Transmitter attachment methods. In: Priede IG, Swift SM, editors. Wildlife telemetry. Sussex: Ellis Horwood. p 135–146.

Angerbjorn A., Becker D. 1992. An automatic location system for wildlife telemetry. In: Priede IG, Swift SM, editors. Wildlife telemetry. Sussex: Ellis Horwood.

Balendonck J, Jansen MB, van den Heuvel CC, Kranenburg RN, Nijenhuis PJ, Wattimena MR. 1990. An automatic Doppler radio tracking system: details of the base and field station. In: Uchiyama A, Amlaner Jr CK, editors. Biotelemetry XI. Yokohama, Japan. p 162–166.

Balwanz M. 1978. Vertical polarized detuning sleeve antenna for a 170 MHz telemetry receiver. In: Klewe HJ, Kimmich HP, editors. Biotelemetry IV. p 21–22.

Beach MH, Storeton-West TJ. 1982. Design considerations and performance checks on a telemetry tag system. In: Cheeseman CL, Mitson RB, editors. Telemetry studies of vertebrates. Symposia of the Zoological Society of London, No. 49. p 31–45.

Bosak A. 1992. Some new design concepts for simple lightweight radio-tracking equipment. In: Priede IG, Swift SM, editors. Wildlife telemetry. Sussex: Ellis Horwood. p 92–97.

Burchard D. 1987. The V-aerial: a navigation aid for tracking down radio equipped wildlife. In: Kimmich HP, Neuman MR, editors. Biotelemetry IX. p 315–318.

Burchard D. 1988a The design of small aerials for close-up tracking. In: Amlaner Jr CJ, editor. Biotelemetry X. Fayetteville AR: Univ of Arkansas Press. p 379–387.

Burchard D. 1988b. Towards higher frequencies in outdoor applications. In: Amlaner Jr CJ, editor. Biotelemetry X. Fayetteville AR. p 57–65.

Dunstan TC. 1977. Receiving systems for studying raptorial birds. In: Long F, editor. 1st International Conference on Wildlife Biotelemetry. Laramie WY. 40–44.

Garrott RA., White GC, Bartmann RM, Weybright DM. 1986. Reflected signal bias in biotelemetry triangulation systems. *J Wildl Manage* 50:747–752.

Gaudin JC, Reudet D, Corbieres JC. 1991. Construction and tuning of "HB9CV" antenna. Office National De La Chasse, Technique No. 28. 4 p.

Gonczi AP. 1992. A method to determine small scale movement of fish by dipole antennas. Manuscript, National Board of Fisheries, Sweden.

Groeneveld WH, de Bakker HV. 1982. Application of an automatic antenna selector. In: Biotelemetry VII. Stanford CA. p 164–167.

Hupp JW, Ratti JT. 1983. A test of radio telemetry triangulation accuracy in heterogeneous environments. In: Pincock DG, editor. 4th International Conference on Wildlife Biotelemetry. Halifax, Nova Scotia. p 31–46.

Jasik H. 1961. Antenna engineering handbook. New York: McGraw-Hill.

Kenward R. 1987. Wildlife radio tagging. London: Academic. 222 p.

Klaus G, Ballisti R, Neukomm PA. 1982. Body mounted antennas: nearfield computations on lossy dielectric bodies. In: Biotelemetry VII. Stanford CA. p 160–163.

Kolz AL. 1991. Supplemental type certificate for Cessna 172 use of wing strut mounted telemetry antennas. FAA Certification #3A12 (patent pending).

Kuechle VB, Reichle RA, Schuster RJ. 1981. Optimum characteristics of receivers for radio tracking. In: Long F, editor. 3rd International Conference on Wildlife Biotelemetry. Laramie WY. p 35–49.

Kuechle VB, Reichle RA, Smith JLD, Schuster RJ. 1983. Directional aircraft antenna system. In: Pincock DG, editor. 4th International Conference on Wildlife Biotelemetry. Halifax, Nova Scotia. p 94–103.

Lancia RA, Dodge WE. 1977. A telemetry system for continuously recording lodge use, nocturnal and subnivean activity of beaver (*Castor canadensis*). In Long F, editor. 1st International Conference on Wildlife Biotelemetry. Laramie WY. p 86–92.

Lemnell PA, Johnsson G, Helmersson H, Holmstrand O, Norling L. 1983. An automatic radio-telemetry system for position determination and data acquisition. In: Pincock DG, editor. 4th International Conference on Wildlife Biotelemetry. Halifax, Nova Scotia. p 76–93.

Macdonald DW, Amlaner Jr CJ. 1980. A practical guide to radio tracking. In: Amlaner Jr CJ, Macdonald DW, editors. A handbook on biotelemetry and radio tracking. Oxford: Pergamon. p 143–159.

Mackay S. 1970. Biomedical telemetry. 2nd edition. New York: Wiley. 540 p.

Mech LD. 1983. Handbook of animal radio-tracking. Minneapolis: Univ of Minnesota Pr. 108 p.

Morris JA. 1992. An easily constructed hand-held direction-finding antenna. In: Priede IG, Swift SM, editors. Wildlife telemetry. Sussex: Ellis Horwood. p 90–91.

Neukomm PA. 1978. Small body-mounted antennas: the influence of the human body on the azimuthal radiation pattern in the frequency range 50 to 1000 MHz. In: Klewe HJ, Kimmich HP, editors. Biotelemetry IV. p 41–44.

Neukomm PA, Klaus G, Ballisti R. 1981. Polarization-transformation effect of the human body and its practical application. In: Sansen W, editor. Biotelemetry VI. Leuven. p 43–46.

Parish T. 1980. A collapsible dipole antenna for radio tracking on 102 MHz. In: Amlaner Jr CJ, Macdonald DW, editors. A handbook on biotelemetry and radio tracking. Oxford: Pergamon. p 263–268.

Priede IG. 1992. Wildlife telemetry: an introduction. In: Priede IG, Swift SM, editors. Wildlife telemetry. Sussex: Ellis Horwood. p 3–25.

Smith RM, Trevor-Deutsch B. 1980. A practical, remotely controlled, portable radio telemetry receiving apparatus. In: Amlaner Jr CJ, Macdonald DW, editors. A handbook on biotelemetry and radio tracking. Oxford: Pergamon. p 269–273.

Taylor KD, Lloyd HG. 1978. The design, construction and use of a radio-tracking system for some British Mammals. *Mammal Rev* 8:117–141.

Uda S, Mushiake Y. 1954. Yagi-Uda antenna. Sendai, Japan: The Research Institute of Electrical Communication. 183 p.

Vreeland RW, Rutkin BB, Yeager CL. 1976. Tuned loop receiving antennas for indoor telemetry. In: Fryer TB, Miller HH, Sandler H, editors. Biotelemetry III. New York: Academic. p 333–336.

White GC, Garrott RA. 1990. Analysis of wildlife radio-tracking data. San Diego CA: Academic. 383 p.

Yasuda H, Inai T. 1990. A review of practical receiving antenna systems used in clinical telemetry. In: Uchiyama A, Amlaner Jr CJ, editors. Biotelemetry XI. Yokohama, Japan. p 341–345.

Chapter 14

Radio-wave propagation effects in avian telemetry

V. B. Kuechle

Radio-wave propagation, which concerns how the radio signal travels between the transmitter and a receiver, varies substantially among different environments. Signal loss between transmitter and receiver due to environmental variables determines whether enough power is available at the receiver to be detectable. The effect of environmental factors, the most difficult variables to predict, can be evaluated via the use of a power budget. This chapter illustrates such effects in various environments, including free space, open fields, and forests. Additional physical factors are discussed, including antenna polarization, signal reflection/refraction, and weather conditions which cause ducting.

Radio-wave propagation concerns how the radio signal travels between a transmitter and a receiver. This chapter will extend that definition somewhat by including a discussion of signal levels at the receiver and transmitter, since doing so will provide an opportunity to use illustrative examples. In this chapter only, propagation at short ranges will be considered; ranges of under several km are generally observed in practice. The signal loss between the transmitter and the receiver determines whether enough power is available at the receiver to be detectable. It is convenient to use a power budget analysis to observe the effect of different influences on the signal power. Using a power budget, we start with transmitter output power and add or subtract signal level effects, resulting in power level at the receiver input. Environmental factors are the most significant and also the most difficult to predict. These effects will be illustrated by considering cases from simple to complex.

The receivers and antennas used today by most telemetry studies are near enough to theoretically optimum design that they are considered to be equal and not a limiting factor. Most receivers have a noise figure of under 2 or 3 decibels (dB) and claim minimum detectable or discernible signal levels (MDS) of about −150 dB relative to 1 milliwatt (dBm). These levels are obtainable in the laboratory. Under field conditions, external acoustic and electrical noise limit the practical signal levels for detection by ear to a signal level of about −140 dBm at the receiver input terminal. If we desire detection by automated electronic means, a signal level of at least −120 dBm is required for reliable detection. Automated electronic detection is required for coded tags or unattended data logging. Normal practice is to use a yagi antenna which will add 9 to 10 dB to the signal level; an omnidirectional antenna such as a dipole will add about 3 dB. Thus we will need at least –150 dBm signal level at the antenna for detection by ear or –130 dBm for detection by electronic means. Digital signal processing (DSP) methods are not considered because they are not being used extensively in animal location applications.

Transmitter output power is the other important part of the power budget and is more difficult to estimate because so many different transmitter types and antenna lengths are used. It is not practical to measure the power output of the transmitters directly since the antennas are rarely matched for optimum power transfer. Also, transmitters are not designed for 50-ohm outputs, which is the normal input impedance for most high frequency test instruments. It is difficult to account for mismatch loss without accurate matching data. A more practical method is to use a reference antenna and measure the induced field at a given distance. These data can then be used to predict the range under various propagation conditions. Using typical transmitters for small birds, power levels of –10 to –30 dBm effective isotropic radiated power (EIRP) are typical. It is also common to specify an induced field at 1 m or at 10 m. From these data, we can make some predictions on range under various propagation conditions.

Case I—free space

Free space propagation is the simplest case and applies only while birds are in flight and when the receiving antenna is elevated. This is the most optimistic calculation, and propagation losses in practice will always be greater than in free space. Even though we are not in free space, atmospheric losses will be ignored. They are very small compared to other losses at the frequency ranges normally used in animal-location studies, especially at the distances we are considering. Most radio-wave-propagation estimation is traceable to optical ray concepts. The limits of free-space propagation under these concepts are shown in Figure 14-1.

Under free-space conditions, propagation is given by the following transmission formula first presented by HT Friis:

$$\frac{p_r}{p_t} = \left(\frac{\lambda}{4\prod d}\right)^2 g_t g_r$$

where:

p_t = power transmitted
p_r = power received
λ = wavelength
d = distance between receiving and transmitting antennas
g_t = transmitter antenna gain
g_r = receiver antenna gain

Equation 14-1.

The Friis transmission formula demonstrates 2 fundamental relationships. First, because wavelength decreases with increasing frequency, the received power also decreases with increasing frequency. While this is important, the choice of frequency is generally made for reasons other than the increased propagation loss due to frequency. The second term of importance is the 1/d term, which is squared in the formula. This is often referred to as the spreading loss and is the result of electromagnetic energy occupying an ever-increasing surface area with increased distance from the source. We can demonstrate with this formula that to double the distance, we must increase the power by a factor of 4 under free space conditions.

Since gains are often given in decibels (dB) and power levels in decibels relative to 1 milliwatt (dBm), it is helpful to rewrite the Friis formula as follows:

$$P_r = (P_t G_t) + G_r + 20 \log\left(\frac{\lambda}{4\prod}\right) - 20 \log d$$

where:

P_r = received power in dBm
$P_b T_b$ = effective transmitted power including antenna
G_r = receiver antenna gain relative to isotropic
d = distance between receiver and transmitter

Equation 14-2.

If we take a typical case as an example:
frequency = 150 MHz; λ = 2 m,
receiver antenna gain = 10 dB,
transmitter equivalent power at 1 m = −40 dBm,
distance d = 10 km,
$P_r = -40 + 10 - 16 - 80$,
$P_r = -126$ dBm which is a power level greater than our −140 dBm threshold.

As stated earlier, this case applies only to reception from a bird in flight to an aircraft or from a bird in flight to an elevated antenna. Figure 14-1 shows the limit on conditions required for free-space transmission. Figures 14-2 and 14-3 show the decrease in signal level versus distance for free-space conditions. Free-space transmission loss is independent of polarization.

Case II—open field sites

Cases in which birds are at or near ground level and observers are also near the ground are more typical cases for study of the effects of pesticides on birds. In this case, ground effects will influence the propagation characteristics. We will first consider open field sites where vegetation contributes little to signal attenuation. This would be true for grassy fields and row crops or small grain agricultural fields. This case is described in the

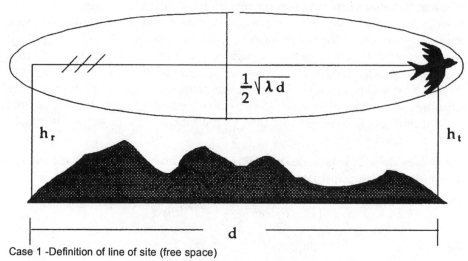

Case 1 -Definition of line of site (free space)

Figure 14-1   Clearance conditions for free space propagation

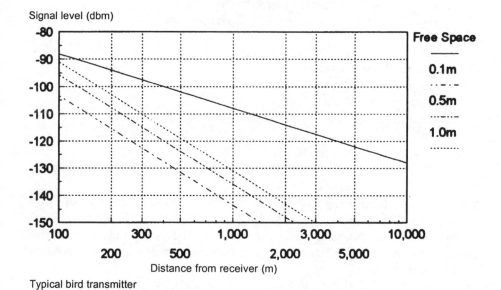

Typical bird transmitter

Figure 14-2   Signal level as a function of height above ground for horizontally polarized antennas

Signal level (dbm)

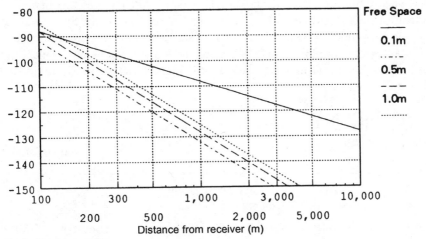

Typical bird transmitter

Figure 14-3   Signal level as a function of height above ground for vertical polarization

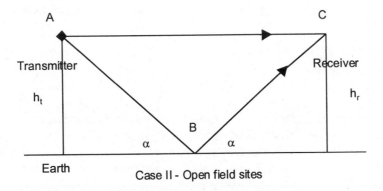

Figure 14-4   Ray tracing for propagation near ground

literature as diffraction conditions (Boithias 1984). Under these conditions, the radio wave has 2 paths: the direct path AC and the reflected path ABC (Figure 14-4). In most of the literature, a smooth earth model is used, and optical ray concepts are the basis of our propagation predictions. If the angle *a* is small, as is the case in bird or animal tracking, we can make some simplifications. Details of these assumptions are available in Bullington (1957). To the formula for free space conditions given previously, we add the following diffraction term:

$$\frac{p_r}{p_t} = \left(\frac{4 \prod h_t h_r}{\lambda d} \right)^2$$

Equation 14-3

where:

h_t = transmitter antenna height above ground
h_r = receiver antenna height above ground.

The other terms are as defined for Case I. Combining the diffraction term with the spreading loss term of Case I yields

$$\frac{p_r}{p_t} = \left(\frac{4 \prod h_t h_r}{\lambda d} \right)^2 \left(\frac{\lambda}{4 \prod d} \right)^2 g_t g_r$$

Equation 14-4.

Combining terms yields the following:

$$\frac{p_r}{p_t} = \left(\frac{h_t h_r}{d^2} \right)^2 g_t g_r$$

Equation 14-5.

Thus the propagation in this case varies to the fourth power of the distance between the transmitter and receiver. We also observe that h_t will go to 0 should the bird be on or near the ground. The formula indicates the power received will also go to 0. We know this is not the case. Using data from Bullington (1947), we use either the actual antenna height or minimum effective height, whichever is greater. Converting the previous formula for use with decibel values yields the following:

$$P_r = (P_t G_t) + G_r + 20 \log(h_t) + 20 \log(h_r) - 40 \log d$$

Equation 14-6.

This loss or gain in the h_t and h_r term is often referred to as antenna-height-gain factor. If the receiving antenna is fixed at 1.5 m, a height common in hand-held tracking applications, we can determine height-gain factors from Figures 14-5 and 14-6. These graphs assume the receiving antenna is at 1.5 m, thus 0 dB should be used as the receiver height gain. If the antenna is higher than 2 m, the graph of Figure 14-6 should be used to determine the height gain. These values can be used in place of the height term in the above equations to determine transmission loss. Using the parameters from the previous example and a horizontally polarized transmitter antenna at 0.1 m results in an antenna

Gain (db)

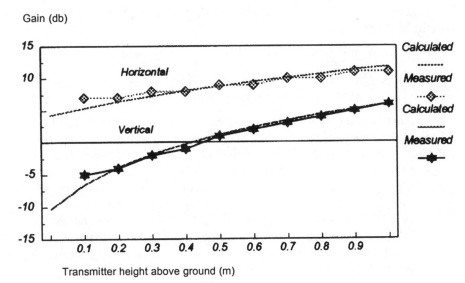

Transmitter height above ground (m)

Freq = 150 MHz
Receiver antenna fixed at 1.5m above ground

Figure 14-5   Height gain as a function transmitter height above ground and polarization

Gain due to height

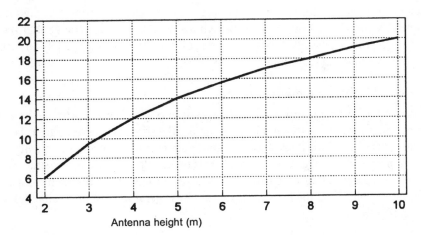

Antenna height (m)

Freq = 150 MHz
Transmitter at 1 m

Figure 14-6   Height gain for horizontally or vertically polarized antennas

height gain of –5 dB and a receiver-antenna height gain of 0 dB because we have normalized the height gain to 1.5 m.

Entering these numbers into Equation 14-6 yields the following:

$$P_r = -20 + 8 - 5 - 160$$

$$= -173 \text{ dBm, which is below the threshold previously set.}$$

These figures also show that with antennas less than 1 wavelength above ground, vertical polarization will give better range than horizontal polarization. Once the transmitter antenna is at least 1 wavelength above ground, signal attenuation is about equal for vertical or horizontal polarization. For antenna heights greater than 1 wavelength above ground, the signal level increases about 6 dB each time the antenna height is doubled (Figure 14-6). This holds true whether the receiving or the transmitting antenna height is increased. The 6-dB increase holds until free space transmission is reached. We have assumed that the distance (d) between antennas is much greater than the product of the antenna and receiver heights. If that were not the case, we would have interference fringes or lobes in the transmission loss.

Soil conditions also have an effect on antenna height gains for heights below 1 wavelength. Good soils (higher conductivity and lower dielectric constant) have higher height gain factors for vertical polarization. Antennas over sea water have the highest height-gain factors. The inverse is true for horizontal polarization with poor soils having higher height gain factors and sea water having the lowest height-gain factor.

Signal attenuation due to vegetation for grass lands and row crops is negligible for frequencies below about 300 MHz. Although many users report less range after vegetation emerges, this is possibly due to changing soil conditions or more likely an increase in noise levels during the summer months.

Case III—forested environments

In forested environments where the vegetation is dense and approaches or exceeds a wavelength in height, we can no longer ignore the effects of the vegetation. The vegetation adds an additional loss factor that is relatively independent of frequency. In addition, propagation takes 2 forms: the first is the direct path through vegetation and the second is by lateral wave transmission where the radio wave skims the tree tops and can be received in either the forest or in the air. This lateral wave is the result of wave refraction (Figure 14-7).

This case can be modeled in the same way we model transmission from freshwater lakes except that the lake is replaced by the forest. Propagation characteristics are determined by the dielectric constant and the conductivity. Since these parameters are determined by moisture and vegetation density, propagation characteristics cannot be predicted as accurately as for lake water. Tamir (1977) gives a range of 1.01 to 1.5 relative to air for the dielectric constant and 10^{-5} to 10^{-3} mho/m for the conductivity.

Figure 14-7 illustrates the possible paths for the forested case. Radio waves can travel by a direct path AC, reflected path ABC, or lateral wave path AEBC. Travel by path AED, as in aircraft tracking, is also possible. These travel paths are defined by optical ray analysis. The reflection and refraction angles are defined by density differences between the forest and the air above.

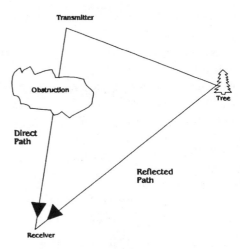

Figure 14-7   Condition for direction-finding errors due to signal reflection or refraction

Attenuation in the forest area is given by L = αd, where α is a constant depending on dielectric constant, frequency, and conductivity. The constant is in dB/m and varies from 0.017 to 1.1 dB/m for the range of dielectric constants and the conductivities given above. Because forests are quite variable, it is difficult to give an accurate number; however, 0.15 dB/m is probably a good midrange value. What we do know is that the loss increases exponentially with distance in the forest but only $1/d^4$ in the air above. For example, at an attenuation of 0.15 dB/m, the signal will be reduced by 150 dB at 1,000 m. Therefore, except for short distances inside the forest, lateral wave transmission will predominate.

In Figure 14-7, this number is multiplied by the path length. In the case of the direct path AC, we can see that if the path length is 1,000 m and the loss is 0.15 dB/m, the signal will decrease 150 dB in 1,000 m. This will place most signals below the threshold for radio location of small radio transmitters. It is easy to see that lateral wave transmission will predominate in most cases because path distance is only from the transmitter to the top of the canopy.

Another phenomena observed in radio wave propagation in forested sites is that horizontal polarization generally gives greater range. Also, even if the transmitter antenna is vertically polarized, the predominant polarization will change to horizontal in the lateral wave. Thus the receiving antenna in forested sites should be horizontally polarized.

Other conditions of interest

There are several other conditions that fall in the propagation category that are of interest. The first of these is the state where the polarization of the transmitter and receiver antenna may not be the same. This is not a problem in hand-held situations because the user can easily change the receiving-antenna orientation. In fact, experienced users usually change the antenna from vertical polarization to horizontal polarization and choose the better method.

If the antenna is fixed, changing the orientation is not as easy, and a polarization mismatch loss may have to be taken into consideration. This loss will be in the range of 5 to 20 dB if the polarization is clearly defined. In most cases

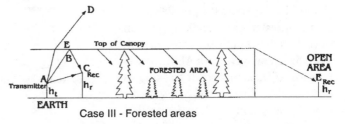

Case III - Forested areas

Figure 14-8   Possible propagation paths in forested sites

there will be a component of horizontal and a component of vertical polarization. Under those conditions, polarization mismatch loss will be in the 4- to 8-dB range.

A second condition is that of signal reflection/refraction. A typical situation is illustrated in Figure 14-8. If there is a significant obstruction in the path of the direct wave, the direct wave will be reduced in level and may in some cases be reduced to an insignificant level. The signal may then travel around the obstruction or may also be reflected off nearby objects such as trees, rocks, etc. The signal reaching the receiving antenna is then the resultant of the direct and reflected/refracted waves. One can easily see that an error in determining the direction of arrival will result. Experienced users can usually detect reflected signals because the pattern does not seem as defined as usual. The remedy is to move to a new location and try again.

Ducting is also a condition sometimes mentioned. This is a situation in which a temperature inversion or other weather condition causes a sharp change in the atmosphere. The radio wave will then be refracted and will stay within the layer, and losses will be greatly reduced. In the frequency and distance ranges we normally encounter, ducting is not an important situation.

References

Boithias L. 1984. Radio wave propagation. New York: McGraw-Hill.

Bullington K. 1947. Radio propagation at frequencies above 30 megacycles. In: Proceedings of the I.R.E. waves and electrons section. New York: Institute of Radio Engineers. p 1122–1137.

Bullington K. 1957. Radio wave propagation fundamentals. *The Bell System Technical Journal* 26(3):593–626.

Tamir T. 1977. Radio wave propagation along mixed paths in forest environments. *IEEE Transactions on Antennas and Propagation* AP-25(4):1–77.

Chapter 15

Fusion of global positioning satellite technology with ARGOS for high-accuracy location and tracking requirements

M. Pike Castles

Continuing advances in miniature global positioning system (GPS) receiver developments have made practical the fusion of high-accuracy GPS location determination with the low-power platform transmitter terminal (PTT) communication link provided by ARGOS. During the past several years, Southwest Research Institute (SwRI) has performed experimental tracking tests using various GPS receivers interfaced to modular PTT designs. Location and tracking accuracies have been evaluated, and a significant effort has been directed toward analysis of bit error rates through ARGOS for various GPS data message formats. This effort has also focused on the optimum design for GPS/PTT interfaces and programmability, emphasizing practical power budgets.

This chapter describes concepts for the employment of GPS-based ARGOS PTT units in the tracking of environmental monitoring assets. Iceberg geolocation, weather balloon tracking, environmental buoys, monitoring, and fixing of mobile or air-deployable remote weather sensors are proposed as candidate missions for the GPS-based PTT units. Results of characteristic tracking experiments are presented for various environmental monitoring scenarios.

One of the technological heroes of this decade is the $12 billion system developed for and controlled by the U.S. Department of Defense. Although GPS is a military system, there are many valuable commercial applications being exploited. Today, manufacturers have recognized these opportunities; and GPS receivers are now being fabricated that are small enough and economical enough to be used by everyone.

As illustrated in Figure 15-1, the GPS constellation, when completed, will consist of 24 satellites, including 3 operational spares, in 12-h orbits 11,000 miles above the earth. The attractive feature of these satellites, which is exploited for tracking purposes, is the fact that the GPS satellites transmit orbital position and precise time based on atomic clocks accurate to within 1 second every 70,000 y. By precisely measuring the time of received signals from several satellites, a GPS receiver can determine its latitude and longitude to within 30 m (typically), and can also calculate altitude, speed, and heading.

Background

To understand the need for GPS/ARGOS satellite tracking, one should examine how GPS can be used with ARGOS and consider the basic differences in the location-determination concepts. Whereas the GPS location concept involves *receiving* satellite signals by the item being tracked, the ARGOS location technique uses a PTT, which *transmits* pulses during a satellite pass. With ARGOS, the location is calculated later using Doppler measurements on the PTT pulses, as measured by the satellite.

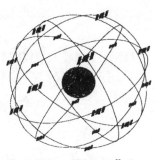

Figure 15-1   GPS constellation

Therefore, as illustrated in Figure 15-2, to use GPS for remote tracking, it is necessary to provide a means to communicate the received GPS location data from the item being tracked to the remote user. This is conveniently accomplished by using a "smart" PTT programmed to format and transmit the GPS location data as part of the PTT message. Notice that the location of the tracked item can now be obtained from a single PTT pulse.

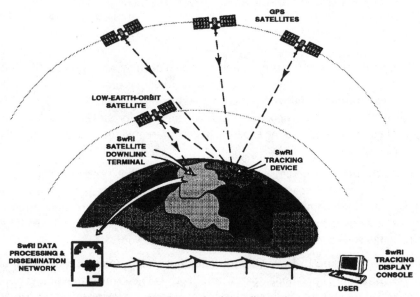

Figure 15-2   GPS/low-earth-orbit satellite tracking concept

GPS/ARGOS developments at Southwest Research Institute

Internal research

During the past several years, SwRI has provided internal research funds to investigate the fusion of GPS technology with low-earth-orbit (LEO) satellites, including ARGOS. The currently available and future planned LEO satellites, such as STARSYS, provide excellent opportunities to communicate GPS-measured data to remote users by using low-power PTTs.

A major portion of the SwRI research effort has been directed toward tracking moving vehicles in the urban environment and small aircraft, both of which are difficult using standard Doppler-processing algorithms. However, these tracking requirements can be successfully met by fusion of GPS and ARGOS techniques. For example, Figure 15-3 shows a vehicle-tracking exercise using GPS in an urban environment. Notice that the location data, which are collected and stored in the PTT every 30 s, easily identify the path of the moving vehicle.

To evaluate the capability to track faster moving platforms, SwRI instrumented a twin-engine aircraft with the smart PTT interfaced to GPS receiver modules. Location data were measured and stored in the PTT during the aircraft flight illustrated in Figure 15-4. This portion of the flight occurred during an ARGOS satellite pass, and the resulting single calculated location using satellite-measured Doppler is identified on the map. The Doppler-calculated location is in error by approximately 120 miles, primarily due to the speed and changing directions of the aircraft. The GPS locations stored during the flight are virtually undisturbed by the aircraft velocity.

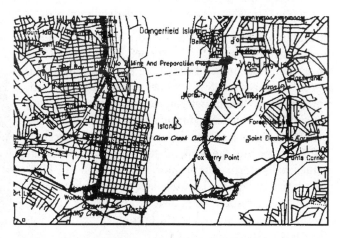

Figure 15-3   Vehicle tracking via GPS (urban environment)

Hardware and software developments

The SwRI experiments confirm that to successfully exploit GPS technology for use in remote tracking requirements, it is necessary to communicate the GPS data through LEO satellites such as ARGOS, using smart PTTs with the following capabilities:

1) interface to GPS receivers,

2) read and store GPS location data,

3) format the GPS data in the PTT message,

4) store satellite ephemeris data to know when to transmit to the satellite.

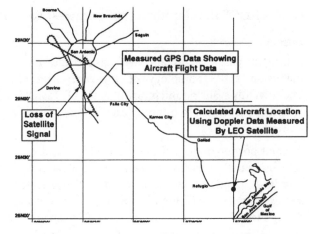

Figure 15-4   Aircraft tracking via GPS

The "Smart" Satellite Transmitter (SST-2A) shown in Figure 15-5 was designed by SwRI to meet these requirements. The unit can be interfaced to a GPS receiver and other sensors, as illustrated conceptually in Figure 15-6. (Technical details are provided in following sections).

In addition to the smart PTT, the user must provide means to decode the PTT message data (whether received from Service ARGOS or a LUT) into usable latitude and longitude information. Again, considerable effort at SwRI has been directed toward optimizing the data throughput to obtain the maximum amount of GPS locations during a single satellite pass. The SwRI GPS/ARGOS tracking system converts the received PTT message into the proper data based on the predetermined coding format for the particular received PTT identification number. The location data are then plotted on a graphic map display, as shown in Figure 15-7, using a world database centered on the latitude/longitude location data received from the PTT.

Figure 15-5   "Smart" PPT (SST-2A)

Technical details

Smart GPS/PTT module

A block diagram of the SwRI GPS/PTT module is provided in Figure 15-8. The PTT microprocessor controls all functions and communicates with the GPS receiver and the external input/output via an RS-232 interface. Prior to PTT deployment, a laptop computer is used to download the GPS/PTT tracking instructions, including

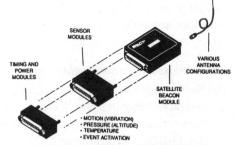

Figure 15-6   Miniature GPS-based beacon

the PTT identification number, repetition rate, required tracking mission lifetime, rate to acquire GPS fixes, satellite visibility data, message format, etc.

An important consideration is the power budget for the tracking mission lifetime. The GPS/PTT real-time clock contains an internal lithium battery, which can provide clock power for approximately 10 y. The clock is also loaded with timing instructions to activate the microprocessor, which controls power to the GPS receiver and the PTT transmitter.

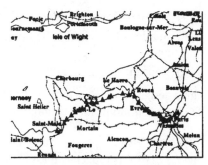

Figure 15-7   Computer-generated   map display showing GPS-measured

Upon microprocessor command, the GPS receiver acquires a location fix, stores the data in memory, and then shuts off to save battery power. The controller reads the time and location coordinates and then determines the proper transmit time based on pre-stored satellite visibility data. (If a GPS fix cannot be acquired within the programmed time interval, the GPS receiver will automatically shut down and retry at the next command time.)

When the satellite becomes visible, the controller loads the GPS data in the predetermined message format and modulates the PTT. It then schedules the next transmit time and shuts power off.

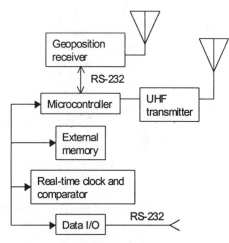

Figure 15-8   "Smart" GPS/PTT block diagram

Message format considerations

The GPS/PTT can store up to 250 time-tagged latitude/longitude/sensor data packets; therefore, it is important to compact the data for optimum transmission throughput using the limited bandwidth in the ARGOS satellites.

The ARGOS message format permits 256 bits of message data to be transmitted in each PTT transmission. Generally, the GPS/PTT tracking-mission requirements will dictate the power-budget constraints and hence the amount of data transmitted and the message format. Parameters that are important include

1) required tracking mission lifetime,
2) battery amp-h rating,
3) required GPS location update rate,

4) length of time allowed for GPS fix before shutdown,

5) number of GPS fixes to be stored and transmitted in each PTT message (256 bits), and

6) transmit repetition rate.

These design parameters must all be taken into consideration to ensure optimum GPS/ PTT data communication through ARGOS.

Data processing

The GPS/PTT compacted message data which is eventually received from Service ARGOS (or from a LUT) must be processed to extract the time-tagged GPS latitude, longitude, and sensor data. Figure 15-9 illustrates a typical output after the PTT message has been processed. In this case, a GPS fix was recorded every 5 min with latitude/longitude accuracy of 1/100 of a minute.

The date/time associated with an individual GPS fix is particularly important for some tracking missions because the specific GPS location point may have been measured and

PTT ID	DATE/TIME (MMDD)/(HHMM)	LATITUDE (sDDMM.mm)	LONGITUDE (sDDDMM.mm)	ALT. (mtr)	SPEED (kt.)	DIRECTION DDD	OTHER SENSOR
1000	0816/1015	+2948.32	+09814.21	05345	241	045	2FF43B
1000	0816/1020	+3002.70	+09828.59	05340	244	045	2FF61A
1000	0816/1025	+3017.13	+09843.02	05340	245	046	2FF61A
1000	0816/1030	+3031.62	+09857.51	05335	246	045	2FF59A
1000	0816/1035	+3046.00	+09911.89	05338	244	044	2FF58B
1000	0816/1040	+3100.32	+09926.21	05336	243	045	2FF58C
1000	0816/1045	+3114.58	+09940.47	05340	242	046	2FF60B
1000	0816/1050	+3128.78	+09954.67	05345	241	045	2FF61A

Figure 15-9   Typical GPS/PTT messages (after processing)

stored in the PTT several hours prior to an opportunity to transmit the data to the ARGOS satellite. The time latency between when the GPS location was stored in the PTT and the time when the satellite was available to receive the transmitted data is of major concern for some tracking missions. Typical values for this time latency are provided in following sections. (Note: The time latency discussed above is different from the ARGOS "data availability," which is the time between reception of PTT data by the satellite and results being available to users on the ARGOS dissemination system.)

Experimental results

Location/track accuracy

Global positioning system location measurements are available in 2 codes: P-code and C/A-code. The P-code (Precise) is encrypted and available only to U.S. government-authorized users. The C/A-code (coarse acquisition) is available to anyone who buys a GPS receiver. Usually the C/A-code provides positioning accuracy to within 30 m. However, the Department of Defense has implemented a system called "selective availability" (S/A) to degrade the C/A-code to prevent it from being used against U.S. military forces. The value of S/A degradation is variable, but typically increases the location error for C/A-code users to 100 m.

The SwRI tracking experiments have confirmed that whenever a C/A-code GPS fix is obtained, the location error is generally less than 100 m, and the location error primarily results from the S/A value at the time of measurement. For this reason, the S/A is continually monitored at SwRI to identify the degree of anticipated location error. (Differential GPS measurements can be performed to remove the major S/A component.)

Figure 15-10 illustrates the variable S/A degradation over a 30-min time period plotted on X–Y coordinates of ± 200 m. During a 12-h period, the S/A typically has the effect shown in Figure 15-11 plotted on ± 100-m coordinates.

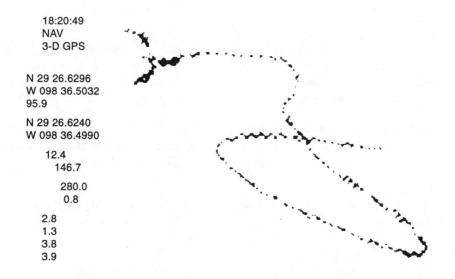

18:20:49
NAV
3-D GPS

N 29 26.6296
W 098 36.5032
95.9

N 29 26.6240
W 098 36.4990

12.4
146.7

280.0
0.8

2.8
1.3
3.8
3.9

Figure 15-10   GPS S/A degradation: 30-minute sample

14:31:01
NAV
3-D GPS

N 29 26.6068
W 098 36.4966
226.9

N 29 26.6240
W 098 36.4990

   32.0
    353.0

    120.6
    0.6

  0.9
  1.1
  1.3
  1.8

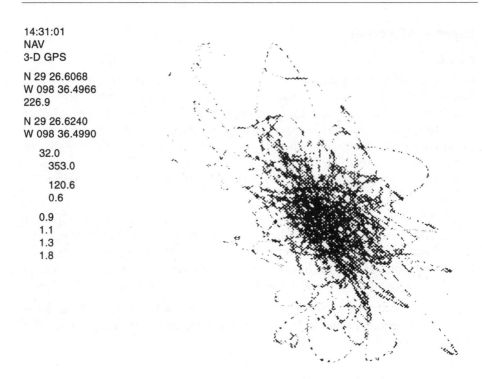

Figure 15-11   GPS S/A degradation : 12 h

Even with S/A, the GPS location accuracy is significantly better than what is obtainable using Doppler processing algorithms and the rapid location-measurement capability, which allows tracking of moving subjects (as shown previously in Figures 15-3 and 15-4) makes GPS/ARGOS attractive for many remote tracking requirements.

Location data error statistics

The other factor that determines the value of GPS fusion with ARGOS is the anticipated bit-error rate and quality of corrupted messages containing GPS data which are sent through ARGOS.

To evaluate this concern, SwRI performed a 5-d tracking experiment covering several hundred miles in which a GPS location was measured and stored in a smart PTT every 6 min. During this test, 1,240 time-tagged GPS latitude/longitude data sets were stored and then transmitted in PTT messages whenever ARGOS satellites were visible. For the total dataset, 87% of the 256-bit messages were received error-free. (Note: The test was performed in the south Texas area where the ARGOS spacecraft utilization is low. The bit-error rate may increase for higher utilization areas.)

Global positioning satellite fix-to-satellite time latency

During the experiment described above, the time latency between when the GPS location was stored in the PTT (every 6 min) and the time when the satellite was available to receive the transmitted data was evaluated. Figure 15-12 provides a histogram of time latency for a sample of 233 datapoints during the experiment. The mean time to transmit the measured and stored GPS data was 3 h and 45 min using only NOAA-11 and NOAA-12.

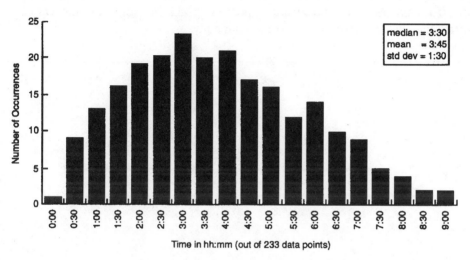

Figure 15-12   Typical histogram of GPS fix-to-satelilite latency

Summary

The fusion of GPS technology with ARGOS enables significant enhancements to tracking and locating missions at the expense of several operational parameters. A summary of the major advantages and disadvantages of this fusion is listed below:

The advantages include the following:

1) high-accuracy location capability,
2) numerous location measurements per h instead of only 1 per satellite pass,
3) the ability to track moving objects,
4) available altitude, speed, and course,
5) only 1 PTT transmission is required for position information on item being tracked,
6) improved satellite utilization with only 1 pulse per PTT location required, and
7) stored GPS location data during times when ARGOS spacecraft is not visible.

The disadvantages include the following:
1) requires smart PTT at additional size and cost,
2) requires that smart PTT include both GPS and ARGOS antennas,
3) requires that GPS antenna have unobstructed view of sky,
4) requires increased power consumption when GPS receiver is on, and
5) requires a programmed tracking mission for optimum power management.

Based on the recent developments and experiments at SwRI, it is anticipated that smart GPS/PTT modules and associated data processing will find numerous applications within the ARGOS tracking and location community.

Chapter 16

Advances in radio tagging and data handling

Robert E. Kenward

Radio tagging is a relatively new technique that is still developing for studying wildlife. This chapter provides a brief assessment of current understanding with recommendations for the future regarding 3 important considerations when radio tags are used for toxicology projects. One important consideration is the choice of an appropriate tagging technique because tags themselves can influence animal behavior and performance. It is also important to collect data that will enhance reliability of survival estimates. Finally, since studies will increasingly test not only whether a given substance has toxic effects, but also whether environmental factors may enhance or diminish those effects, it is important to use robust and cost-effective techniques for analyzing home ranges and habitat use.

Assessment of tagging techniques

The first radio tags and receiving systems used widely for wildlife were developed during the 1960s. Most techniques used for attaching tags to birds were developed by the early 1980s. Experience has, therefore, accumulated for at least 10 y on implantation, on tag attachment to legs, tails, necks, beaks, and patagia, and on backpacks attached either with harnesses or with glue (occasionally combined with sutures). Descriptions of these mounting methods are reviewed in Cochran (1980), Mech (1983), Kenward (1987), Anderka and Angehrn (1992), Samuel and Fuller (1994). It is, therefore, relatively easy to compare the time or skill needed to attach tags in different ways, at least in qualitative terms (Table 16-1). It is also possible to assess the rate of tag loss when using different attachment techniques, thus taking into account the relatively rapid detachment of glue-mounted tags, the molting of tags fastened to tail feathers, or the loss of antennas, which tends to disable tags on bird legs.

Many authors have published comparisons of different mounting methods, and a few studies have compared tagged with untagged individuals. Some studies have shown adverse effects of particular tagging methods on behavior, survival, or breeding success; others have sought adverse effects and found none. A summary of the findings is complicated by variation between techniques in the frequency of testing for adverse effects. Therefore, Table 16-1 provides not only an assessment of risk from an attachment technique, but also indicates the reliance to be placed on the assessment. For example, there

Table 16-1   *A qualitative assessment of disadvantages for different radio tag attachment techniques. The rating decreases to* —— *when there is most disadvantage.*

Technique	Handling time	Skill requirement	Rate of loss	Risk of effect	Other limitations
Implant	——	——-	-	—[1]	Poor signal range, invasive
Patagial	-	—	—	—[1]	Only for large, slow fliers
Beak-mount	-	—	—	—[0]	Only for medium to large birds
Leg-mount	-	-	——	﹍[1]	Not for very small birds
Tail-mount	——	—	—	﹍[1]	Not for very small birds
Necklace	-	-	-.	—[1]	Behavior, icing, crop-shape
Backpack: glue	-	—	——	—[1]	Detachment rate varies greatly
body-loops	——	——-	-	——[2]	Should have case-by-case tests
thigh-loops	-	—	?	?[0]	Needs further testing

[0] Risk ratings are based solely on unpublished data
[1] Risk ratings are based on 1 to 10 published articles
[2] Risk ratings are based on > 10 published articles

are many papers on effects of backpacks attached by body-loop harnesses, but fewer than 10 on necklaces, and none as yet on backpacks mounted with thigh loops. Considering that hundreds of studies have used radio tagging, it is rather disappointing how few have taken the fundamental step of testing for effects on their study animals.

The most satisfactory techniques in Table 16-1 are those with fewest – signs. As all techniques for which there is much information have some disadvantages, the choice of technique is influenced by the type of study. For example, although leg-mounted tags are prone to premature signal loss through antenna damage, the ease of using them safely makes them attractive for some short-term work. For longer studies, neck-mounted tags have proved to have no disadvantages in several studies of game birds, but these tags seem to affect behavior (especially flight activity) in some passerines. For maximum life without the relatively short range of implanted tags, backpacks on body-loop harnesses are the only well-documented attachment. Although some projects have reported reduced survival of birds with body harnesses, other careful comparisons have detected no adverse effects, especially for young altricial species tagged before leaving the nest. Harnesses may, therefore, be safe if attached very carefully at a noncritical time of life. The new backpack attachment (Rappole 1991; Poché et al. this volume), with a lateral loop around each thigh, looks promising but requires further assessment before it can be considered safe for long-term tagging.

In reality, the attachment of a radio tag must always have some adverse effect, if only because a bird uses a little more energy to carry the extra weight (Pennycuick and Fuller 1987; Pennycuick et al. 1989). If no adverse effect is found, it probably only means that the adverse effect was too slight to be detected in the study concerned. What happens, however, if there is a detectable adverse effect? Such an effect may be acceptable in an experimental study, if it has no serious implication for animal welfare, provided that its impact would be similar for both treatment and control groups. On the other hand, an effect would be unacceptable if it interacts with the experimental treatment and thus has a disproportionate impact on the experimental group. An example would be a tagging technique that changes feeding habits so that contaminated prey is eaten less often by tagged animals than by those without tags.

The possibility of such interactions makes it especially important to discuss a proposed study thoroughly with tag manufacturers. Tag makers can often provide more information than a literature survey because they tend to collect unpublished observations such as, "This wasn't worth mentioning in our study, but ...," as well as, "This didn't work at all." Another good reason for thorough advance discussion is that no manufacturer carries large stocks of tags suitable for each possible study, because the demands of different projects are too diverse. For toxicology studies in particular, where large numbers of tags may be needed, it behooves project leaders to organize tag supplies several months in advance.

An improvement for survival estimates

Techniques for estimating survival of radio-tagged animals are reviewed adequately elsewhere (White and Garrott 1990; McDonald et al. this volume). There are 2 main approaches, both developed originally for human epidemiology. The interval-product approach (Kaplan and Meier 1958; Trent and Rongstad 1974; Heisey and Fuller 1985; Pollock et al. 1989) estimates survival over a long period as the product of records for many short intervals between survival checks. Separating the intervals makes allowance for animals that enter the tagged population at different times or that leave a monitored population through signal loss as well as death. This approach produces a simple survival estimate that is suitable for comparing 2 categories of animals at a time, e.g., in treatment and control areas. The second approach can accommodate a number of covariates that describe each individual by using a regression approach to construct multivariate hazard functions (Cox 1972; Kalbfleisch and Prentice 1980; Sievert and Keith 1985). A survival investigation could, for instance, take account of possible age, sex, or weight differences between animals at treatment and control sites. The drawback to this approach is greater restriction on entry and exit conditions.

In both cases, there is a problem if signals from many tags are lost during a study. If signals are lost for reasons which do not affect the survival of the wearers (e.g., because of component failures in tags or molting of attachment feathers), then survival estimates can include the presence of these censored animals until their loss and simply omit them thereafter. If, however, some censored animals die when their signals are lost, perhaps because they are eaten by a predator that also destroyed the tag, then the censoring overestimates survival. Unjustified censoring could be a particular problem in toxicology projects if animals killed by a chemical treatment are rapidly scavenged and tags are thereby destroyed.

This problem can be solved for interval-product survival estimates, provided that reasonable numbers of tagged animals can be identified independently of the tagging (e.g., by re-trapping or band recovery) after the period of survival monitoring. It must be possible also to separate tags that disappeared inexplicably from those that were lost for known reasons, including being molted or showing marked signal changes (indicating failure) just before they were lost. For interval-product estimates, the function $(1-d_i/r_i)$ is used to estimate survival during each interval i, in which d_i deaths were recorded from r_i birds at risk. In this case, if some of u_i birds lost without explanation during interval i had been destroyed with their tags, the true survival would have been less than $(1-d_i/r_i)$. Unless all u_i tags had been destroyed with their wearers, the survival would have been greater than $(1- (d_i + u_i)/(r_i + u_i))$ as well. These 2 functions represent maximum and minimum estimates of the true survival. Given the destruction of some tags with their wearers, such that the recovery rate for u animals with unexplained tag losses (f_u) is less than the recovery rate (f_e) for e animals with explained tag losses, then the function $\zeta = (1-f_u/f_e)$ can be used to interpolate a best estimate of the survivorship as

$$\exp (\ln s_{max} - \zeta.(\ln s_{max} - \ln s_{min})) = (1 - d_i/r_i)(r_i/(r_i + u_i))^\zeta \qquad \text{Equation 16-1.}$$

For details, with confidence limit estimates as an appendix by NJ Aebischer, see Kenward (1993).

More work is now needed to show how such a correction factor can best be applied in hazard modeling. In the meantime, however, if some tags are lost inexplicably during trials, it would be wise to run hazard models twice: once with the assumption that lost signals represent birds that died at the time of loss and once without this assumption. It is also worth taking steps to detect when signal loss is likely due to tag failure, e.g., by recording signal repeat rates at each survival check, and to gather data (such as by re-trapping) which might show that lost signals did not represent dead animals. Future studies may be able to avoid problems caused by unexplained tag loss, partly because of increasing reliability of tags and partly through using automated systems to detect deaths and alert observers before scavengers have time to find carcasses. Even without automated systems, sensors in tags including thermistors that indicate death through cooling and mercury switches that show lack of movement can help to detect the death of a tagged animal with minimum delay. There is much scope for the development of intelligent logging systems to detect deaths or tag loss and to provide an immediate alarm by interpreting signals from simple tags, thus avoiding the added cost and weight of tags with built-in mortality sensors.

Assessments of home range and habitat use

Sometimes it is not possible to arrange replicated survival assessments in experimental and control areas. In this case, it may nonetheless be possible to show that survival is related to an individual's exposure. Such evidence can also strengthen the evidence of a toxic effect where treatment and control areas were inadequately paired, such that observed survival differences could have reflected site differences independent of the treatment.

One measure of exposure would be the area of treatment within an individual's home range, e.g., the number of sprayed fields, representing the availability of contamination. Other perhaps better exposure estimates would be the extent to which an animal used the sprayed fields, as shown by the pattern of radio locations within its home range, and the use of such fields relative to their availability. If home ranges and habitat use are to be investigated, very careful planning is required.

Especially in studies where many animals are tagged, it behooves researchers to discover at an early stage in a project how many datapoints are needed from each individual. Many studies have now shown, for example, that 30 radio locations taken over 10 to 15 d can provide a range outline that does not increase greatly in size as further locations are added (Harris et al. 1990; Kenward 1987, 1992). There is little point in collecting hundreds of radio locations for each individual if 30 suffice. If few locations are needed from each individual, many individuals can be monitored at a time, especially in relatively dense populations, without the need for automatic location systems. Analyses can then be based on one estimate of range size or habitat use for each animal, which avoids the dubious assumptions of data independence or distribution in analyses based on individual radio locations. In various studies, 1 mobile tracker has collected data on up to 16 raptors, 28 squirrels, and 60 pheasants at a time. The assessment of habitat availability and use occurs at 2 levels. One level reflects how each range is placed in the study area. The other level reflects how the animal moved within that range. Although several types of outlines may be used to denote the range, convex polygons are probably most common (Harris et al. 1990). Use of habitats or treatment areas can be analyzed in several ways (Alldredge and Ratti 1986, 1992; White and Garrott 1990). Compositional analysis may prove the most robust yet flexible approach (Aebischer et al. 1993). It follows the rationale of using animals, not radio locations, as sample units and also avoids the problem that habitat proportions are not independent (because they sum to unity). Use of each habitat U_i is expressed relative to one of the other habitats U_j, as a logratio, $\ln(U_i/U_j)$; with 2 habitats this generalizes to $\ln(p/q)$. Similarly, availability is $\ln(A_i/A_j)$. These ratios are symmetrical about 0, and the difference between them indicates preference between the 2 habitats. Summing logratios across individuals gives a matrix of means and variances for each habitat pair that can be used to estimate Wilk's lambda as an overall test for evidence of significant habitat preference, with extension to more sophisticated multivariate tests. Student's t can be used to test whether the preference differs significantly from 0 for each habitat pair, and summation of t values for each habitat allows ranking of the habitats by preference. Estimation of areal availability and use has been simplified by GIS systems and the Ranges IV package for radio-location data (Kenward 1990), but there is still a need for software to simplify the ranking and testing of preferences.

Conclusions

This chapter provides suggestions, based on recent studies, for choosing how to attach tags, for improving accuracy in survival analyses, and for enhanced rigor with reduced effort in location analyses. Opportunities for future progress are identified. There is a need for more data not just on the safety of new tagging techniques, but also on possible interactions between tagging techniques and experimental treatments. More work is needed on techniques to correct for tag disappearance in survival analyses and on systems that can detect death and tag loss automatically. There is also scope for more software to simplify habitat-based analysis of exposure to toxins.

More immediately, it is worth considering all 3 areas carefully before starting a study, ideally by using them as the basis for pilot work. Preliminary tagging tests are essential if possible adverse effects are to be detected in time to modify the techniques. A pilot study is also advisable for maximizing the efficiency of data collection. In survival studies, a pilot study can show how mortality may best be detected, how frequently animals need to be checked, whether loss of signals is likely to be frequent, and if so, how a correction might be applied. In studies which involve home-range assessment, preliminary work will show how few locations are needed from each animal, and thus how many tagged animals can be monitored per person in the field. In a study of many tagged animals, it is prudent to become familiar in advance not only with the radio-tracking hardware, but also with the limitations of analysis software and of the project's human element.

References

Aebischer NJ, Robertson PA, Kenward RE. 1993. Compositional analysis of habitat use from animal radio-tracking data. *Ecology* 74(5):1313–1325.

Alldredge RJ, Ratti JT. 1986. Comparisons of some statistical techniques for analysis of resource selection. *J Wildl Manage* 50:157–165.

Alldredge RJ, Ratti JT. 1992. Further comparison of some statistical techniques for analysis of resource selection. *J Wildl Manage* 56:1–9.

Anderka FW, Angehrn P. 1992. Transmitter attachment methods. In: Priede IG, Swift SM, editors. Wildlife telemetry—remote monitoring and tracking of animals Chichester England: Ellis Horwood. p 135–146.

Cochran WW. 1980. Wildlife telemetry. In: Wildlife management techniques manual. Washington DC: Wildlife Society. p 507–520.

Cox DR. 1972. Regression models and life tables. *J R. Statistical Soc B* 34:187–220.

Harris S, Cresswell WJ, Forde PG, Trewella WJ, Woollard T, Wray S. 1990. Home-range analysis using radio-tracking data—a review of problems and techniques particularly as applied to the study of mammals. *Mammal Rev* 20:97–123.

Heisey DM, Fuller TK. 1985. Evaluation of survival and cause-specific mortality rates using telemetry data. *J Wildl Manage* 49:668–674.

Kalbfleisch JD, Prentice RL. 1980. The statistical analysis of failure time data. New York: John Wiley and Sons. 321 p.

Kaplan EL, Meier P. 1958. Nonparametric estimation from incomplete observations. *J Am Statistical Assoc* 53:457–481.

Kenward RE. 1987. Wildlife radio tagging. New York: Academic Press. 222 p.

Kenward RE. 1990. Ranges IV. Software for analysing animal location data. Wareham UK: Institute of Terrestrial Ecology.

Kenward RE. 1992. Quantity versus quality: programmed collection and analysis of radio-tracking data. In: Priede IG, Swift SM, editors. Wildlife telemetry—remote monitoring and tracking of animals. Chichester, England: Ellis Horwood. p 231–246.

Kenward RE. 1993. Modelling raptor populations: to ring or to radio-tag? In: LeBreton JD, North PM, editors. Advances in life sciences: marked individuals in the study of bird population. Basle, Switzerland: Birkhaeuser.

Mech LD. 1983. Handbook of animal radio-tracking. Minneapolis MN: University of Minnesota Press. 107 p.

Pennycuick CJ, Fuller MR. 1987. Considerations of effects of radio-transmitters on bird flight. *Proc Int Symp Biotelemetry* 9:327–330.

Pennycuick CJ, Fuller MR, McAllister L. 1989. Climbing performance of Harris' hawks *(Parabuteo unicinctus)* with added load: implications for muscle mechanics and for radiotracking. *J Exp Biol* 142:17–29.

Pollock KH, Winterstein SR, Bunck CM, Curtiss PD. 1989. Survival analysis in telemetry studies: the staggered entry design. *J Wildl Manage* 53:7–14.

Rappole JH, Tipton AR. 1991. New harness design for attachment of radio transmitters to small passerines. *J Field Ornithol* 62(3):335–337.

Samuel MD, Fuller MR. 1994. Wildlife radio telemetry. In: Bookout F, editor. Wildlife management techniques manual. 5th edition. Washington DC: The Wildlife Society. p 370–418.

Sievert PR, Keith LB. 1985. Survival of snowshoe hares at a geographic range boundary. *J Wildl Manage* 49:854–866.

Trent TT, Rongstad OJ. 1974. Home range and survival of cottontail rabbits in southwestern Wisconsin. *J Wildl Manage* 38:459–472.

White GC, Garrott RA. 1990. Analysis of wildlife radio-tracking data. New York: Academic Press. 383 p.

Chapter 17

Automatic positioning and data collection systems for desktop tracking of wildlife

Jim Lotimer

Contemporary automatic positioning and data logging systems used for wildlife radiotelemetry studies are reviewed. A brief historical perspective on early systems is provided along with a discussion of their limitations and problems. Modern equipment is functionally discussed with practical examples reviewed. A discussion of automatic data logging of temperature, electromyograms, presence and absence, movement, activity patterns, cellular locations, bearings, remote system control, and data retrieval is included. Features and limitations of these systems are presented with reference to specific system design issues such as validation of signals, coding schemes, radio-frequency interference, noise, time to acquisition, numbers of individuals, size, life, and range. Lastly, this chapter presents a brief overview of future automatic system expectations stressing developments that will extend the limits of present technologies.

Automatic equipment for positioning, monitoring, and tracking wildlife has been in use for many years. Today, both ground-based and satellite-based systems offer full automation of the data collection process. Systems available include

1) positioning from fixed towers (Doppler, Signal Strength, etc.),
2) retransmitted global positioning satellite (GPS),
3) stored GPS,
4) inverse Loran C or tracker mode local area systems,
5) time-of-arrival systems,
6) ARGOS satellite systems and platforms,
7) ORBCOMM satellite (1996),
8) GEOSTAR satellite (1996),
9) automatic data collection receivers, and
10) cellular positioning systems (using antenna radiation patterns and signal strength analysis).

When examined within the context of providing equipment for small birds, some of the above methods lack suitable miniature-transmitter hardware and are considered impractical. For use on small animals, portable, automatic wildlife telemetry systems are forced to use more conventional technologies. The following discussion outlines contemporary automated wildlife radiotelemetry systems, their performance, limitations, advantages, and disadvantages as they relate to small avian species.

Background

Many wildlife radiotelemetry studies still use trained radio operators for detecting, positioning, and data recording of transmitted radio signals. This procedure, by modern terms, is inefficient, expensive (labor intensive), error prone, and often produces small databases not capable of providing detailed analysis. While equipment exists today that will fully automate the process of data gathering and analyzing wildlife radiotelemetry signals, its use has been limited. Finite funds for up-front capital expenditures, misconceptions of capabilities, inertia, and lack of education on automated techniques stand out as the greatest restrictive factors to a more wide spread use of these systems.

For birds, the use of automated systems has not been fully exploited because techniques used in the design and manufacture of avian transmitters has not, fundamentally, changed in 20 y. Transmitters have been made smaller and lighter, and receiver features have been increased; however, the use of periodic on/off keying of carrier waves transformed (in the receiver) to coherent audible signals still represents the majority of equipment available for field studies. This system architecture was chosen for simplicity of transmitter designs and has one other primary advantage: the ability of the human auditory system to sort coherent signals below background noise levels. This advantage ultimately shows up in greater system ranges, perhaps the most pursued feature of wildlife radiotelemetry systems.

Unfortunately, this architecture is not optimized for automatic data collection. As a result, attempts in the past to force the designs to comply with automated desires has led many a frustrated biologist to return to the hand-held yagi antenna and the "beep, beep, beep" human data logger. Some of these frustrations are from equipment failures (mechanical and performance), but more often quality, reliability, and interpretation of data are poor and highly suspect due to the presence of unwanted noise.

Noise fundamentals

Whether you are trying to locate or analyze signals (manually or automatically), your greatest single source of problems is likely to be noise, or more properly, the ratio of signal power to noise power (SNR) in your particular environment. Under ideal conditions (on the tundra perhaps, or inside a shielded chamber where the noise energy may be as low as −155 dBm), the receiver's threshold capabilities become the limiting factor for reception of the signal. Under these conditions, receiver-data loggers are available that will allow users to detect, by ear, pulsed signals whose received power is somewhat less than −145 dBm and will automatically acquire, measure, and data log signals at a level of around −135 dBm.

In the case where the noise floor (average amount of random electromagnetic energy in the local environment) is the dominant factor (noise floors of around –130 dBm or above), the user can hear an uncoded signal that is 12 dB below the noise floor while the signal must be 6 dB above the noise floor for electronic recognition. For example, if the noise floor is –130 dBm, the user will be able to hear a signal of –142 dBm and data log a signal of –124 dBm. However, if the noise floor is –142 dBm (near the receiver's threshold), audio signal detection will still be around –145 dBm (limited by the receiver's threshold), and data logging will be achieved at –135 dBm (again limited by the receiver's threshold). With respect to audio signals, this specification is similar for all conventional wildlife receivers. However, with respect to data logged signals, receiver specifications can vary widely from manufacturer to manufacturer. In all cases and in non-ideal environments, minimum discernible signal levels will rise with the noise floor.

Even if the SNR is adequate, automatic systems may still need to be compensated in order to deal properly with noise effects. High absolute levels of noise can saturate the receiver, reducing the effective SNR, and can prevent signal acquisition by overburdening the processor (data detector). Interestingly, the ear is subject to similar constraints. Thus, the first line of defense against noise is to reduce the receiver gain. Some modern data logging programs employ 2 complementary gain reduction strategies which automatically adapt to time- and frequency-varying noise levels.

Some forms of noise are naturally bursty, like mobile voiceband messages or satellite transmissions. Here the best remedy for the automatic receiver is to attempt to reject signals with inappropriate time signatures. Setting window boundaries tightly around the expected pulse interval of the transmitter will help prevent bursts of noise from being reported as signals. It will also help relieve congestion in the processor, since invalid events take less time to process than legitimate ones.

Both gain-reduction and time-interval filtering have limited usefulness if the dominant noise source is impulsive. Engine noise of all kinds falls into this category. Impulsive noise is characterized by repetitive, but typically very narrow, pulses—each with sufficient peak power to be heard or recognized by the receiver even though the average noise power may be well below the level of the desired signal. In such cases automatic systems can distinguish signals from noise on the basis of pulse duration.

Considerations for a practical automatic system

While site-and-application specific analysis should be conducted prior to choosing a system architecture for your automatic requirements, some issues that are common include

- providing sufficient transmitter output power (range) while providing adequate operational longevity to meet study objectives,
- selecting and using a variety of coding formats in transmitters to distinguish individuals,
- placing radio receivers with automatic monitoring and data storage capabilities in environmentally hostile and remote sites for extended periods,

- retrieving data and controlling the monitoring stations via remote communications devices,
- providing a sufficient choice of remote communications options to overcome limitations imposed by geographic isolation, and
- positioning both automatically and from vehicles.

Range

To conform to proper scientific standards the effective radiated power (ERP) (the real RF power radiated from the radio tag) of a transmitter should be expressed in engineering units such as dBm. However, it is often expressed by some manufacturers as a function of a maximum range (kilometers or miles), making system analysis impossible.

In free space, a 6-dB increase in ERP of a transmitter will double its range, but it can also represent an increase in the input (battery) power of up to 16 X, depending somewhat on the transmitter antenna efficiency. Needless to say, this has a significant negative impact on the operational life and battery requirements to power the device. To lessen the impact of the input energy demands, the duty cycle (a combination of the pulse interval [the number of pulses per minute] and pulse duration [width of the pulse]) of the unit can be decreased, but it will be at the expense of detectability when audio tracking. This is not necessarily the case when using an automatic system, and significant energy savings can be achieved through the reduction of duty cycle. This energy savings can be used to increase output power (ERP), thus enhancing the range of an automatic system.

Range of reception of a radio signal is dependent upon a number of factors, some of which are

- noise floor level (dBm) (bandwidth dependent),
- receiver sensitivity (dBm),
- antenna gain and height (dBd),
- transmitted power (ERP in dBm),
- propagation loss (dB),
- signal to noise ratio (dB), and
- frequency used (MHz).

The noise floor will have the greatest negative effect on range in an automatic system. Modern receiver/data loggers have a signal capture level of –135 dBm (with an SNR of 6 dB). In many cases, the noise floor will be much greater than this level and will thus be the limiting factor in signal reception range.

To demonstrate the ideas discussed above, we will work through an example system requirement for range versus power in a free-space environment.

Required data logging range = 25 miles (42,000 m), air to air

Transmission frequency, 166 MHz

1) Signal losses—examination of free space (line of sight) signal losses shows that at 166 MHz, the propagation loss is:

   Loss   = $22 + 20 \log \underline{d}$

   Since

   $$= \text{wavelength in meters} = 1.8 \text{ m at } 166 \text{ MHz}$$

   and d =   range in meters = 42,000

   Therefore, the propagation loss is calculated as 109.3 dB (approximately 110 dB).

2) Knowledge of the noise floor levels and the antenna gains allows us to calculate the required transmitter power (these parameters are measured at the site of interest):

   Consider a system where the average noise floor is –120 dBm (a typical level in an urban North American environment) and the receiving antenna has a gain of 6 dB. Since the noise floor is above the receiver sensitivity, it alone will be the limiting factor when calculating the transmitter ERP required.

Start with	Noise level (system sensitivity)	= –120 dBm
add	Propagation loss	= 110 dB
subtract	Receiver antenna gain	= 6 dB
Transmitter power required (ERP) for received signal power equal to the noise floor		= –16 dBm
Transmitter power required (ERP) for audio detection of the signal		= –28 dBm
Transmitter power required (ERP) for data logging detection of the signal		= –10 dBm

Note again that a typical wildlife beeper can be audibly detected 12 dB below the noise floor level, but a signal level 6 dB above the noise is required for good data collection.

One can quickly see that the local noise floor will usually determine the range of the system. It is not uncommon for some local noise floors to be at –110 dBm. This would require a transmitter signal strength of 0 dBm for automatic data logging in the example case. The above analyses are provided for a theoretical look and must be used with caution where local propagation or antenna efficiencies vary considerably.

It is important to note that several components within the system, other than output power, will impact on range and detectability of signal. These include sensitivity of the receiver in conjunction with noise floors at the operating frequency within the locality where monitoring takes place; gain and height of the receiving and transmitting antennas; and the degree of hearing impairment of the receiving equipment operator (audio tracking). For example, replacing a 4-element yagi (6-dBd gain) with a 9-element yagi

(11.1-dBd gain) as the receiving antenna will nearly double the range of the system (all other elements of the system remaining the same). Also, increasing the height of the receiving antenna for each 12 feet above the first 10 feet will (theoretically) double the range (i.e., 6 dB), assuming cable losses are negligible.

The effects of human hearing on signal detection are addressed by Kolz and Johnson (1981).

Coding formats

Requirements to mark many birds in conjunction with a restricted number of frequencies dictate that some form of encoding be used to identify individuals. Not only does coding increase the number of individually identifiable transmitters placed at a particular frequency, it also has the effect of reducing the number of frequencies required to mark a given number of animals and of reducing the number of frequencies needed to be scanned during tracking sessions (i.e., automatic, remote, and aircraft tracking). The latter reduces the probability of missing a bird as a result of long scanning cycles relative to the speed of the tracking vehicle. The same holds true for a transmitter moving through a fixed receiving site.

A number of coding schemes are available, including the following: pulse repetition coding (bursts of multiple pulses), pulse rate coding, pulse width coding, on-off keyed Manchester coding, and specially designed codes sets for wildlife applications. Power efficiency of the coding scheme for the best life and range is of ultimate importance in a wildlife system and is optimized in contemporary designs. Reduced range, as a result of automatic detection sensitivity losses and subsequent code discrimination at fixed sites, may be partially or fully compensated in various ways including increased ERP. Code formats which provide distinguishability by ear may be preferred where identification of individuals during audio tracking sessions is important and where flight transect distances are required to be maximized to effectively cover large geographic areas. Interestingly, it would appear that audible code patterns increase detectability by ear of signal from noise, thereby increasing range. This phenomenon was reported for single versus double pulse coded transmitters of equal output power in a fish tracking program in British Columbia (Mark Beere, personal communication).

The presence of a large number of coded transmitters at a particular receiving site increases the risk of code collisions and errors in identification. Code schemes should be selected so as to minimize the probability of collision errors while optimizing the number of individuals that can be placed on a single frequency. Also, mathematical error-correction capabilities are available to further increase the security of data. The potential for use in bird or waterfowl studies requires that the selected coding format be expandable to accommodate a large number of birds.

Automatic monitoring in remote, hostile environments

Monitoring of staging and wintering areas, for example, will require deployment of self-supporting automatic data logging systems over a wide geographical area. Units will be subjected to a variety of hostile elements. Consequently, they must be able to operate throughout and withstand (or be housed in containers able to withstand) wide fluctuations in temperature, rain and snowfall, icing, high winds, humidity, and salt corrosion. Lightning suppressors may be required on all fixed installations.

Data storage capabilities must be sufficient to sustain data-collection integrity during a temporary communications failure, should it occur. Likewise, the system should have a sufficient internal power supply to be able to continue functioning in the event of an external power failure. Memory protection through board level battery backup is necessary in the event of total failure of external and internal power supplies.

Remote communications, interrogation, and control

If the geographical scope of the project is such that stationary field systems must be frequently interrogated for downloading of data (for flight planning, etc.) or to change receiver parameters, central sites with appropriate communications support (i.e., direct phone or cellular phone) are available and can be provided. Where cellular phone coverage is not available and direct phone link is impractical, the capability to link via direct radio modem or satellite is ready. Accidental or deliberate access to the receiving system via the communication links by other than project personnel must be protected through the use of access codes.

Automatic triangulation and position

Automatic direction finding (ADF) and subsequent automatic positioning systems are an exciting application of modern receiving systems. An APS automates the manual tracking and data-collection process, thereby allowing continuous and accurate data-collection 24 h/d, 7 d/week. This dramatically improves the accuracy of habitat studies since many more datapoints are available for statistical interrogation with minimal variance. As a minimum, an automated system is able to match the accuracy of manual triangulation, yet improves on data volume. Mathematical curve fitting and interpolation algorithms, in fact, improve resolution over manual methods. An APS can be realized using a wide range of methods, including satellite techniques such as GPS, retransmitted GPS and ARGOS; ground-based navigational techniques such as Loran-C, retransmitted Loran-C and Inverse Loran; and ADF techniques using Doppler arrays, mechanically scanned arrays (MSAs), phased arrays, and switched yagi arrays (SYAs). To isolate the technique best suited to a given application, it is necessary to consider factors such as transmitter size, weight, and lifetime; sensor requirements; subject movements; terrain; canopy; climate; number of subjects; location accuracy; power availability; processing requirements; and delivery schedules and budget.

Review of the system requirements for an APS eliminates some APS implementation options immediately. As mentioned in the introduction, GPS, retransmitted GPS, Loran-C, inverse Loran, retransmitted Loran-C, and ARGOS systems can be eliminated for applications on small birds due to size.

A more basic APS system, utilizing conventional wildlife technologies, consists of 3 ADF stations. Although 2 unique intersecting bearings are all that are needed for triangulation, the error polygon associated with just 2 bearings can increase dramatically as the transmitter approaches the area between 2 stations. By providing a third bearing from a station located a reasonable distance away from the collinear path between the 2 original stations, this error can be minimized. The accuracy of an individual ADF station is dependent on several factors. First, accuracy is improved by lowering the variance of the ADF bearings as measured by using a stationary transmitter. Secondly, accuracy is improved by decreasing the time necessary for the ADF station to acquire a bearing relative to the average speed of the animal; however, a reduction in this delay usually occurs at the expense of the bearing variance. Third, accuracy is improved by synchronizing readings made by all stations on the same transmitter, especially if the animal is moving.

With ADF and APS, data can be collected and processed in many different ways. It can be stored at each ADF station for several days until manually collected and assimilated in the research laboratory. Alternatively, the data could be downloaded via radio links in preparation for post-processing. However, the most convenient method would be to install dedicated radio links to allow real time data collection and display of animal coordinates on a GIS computer display.

There are several methods by which ADF can be performed, each having its own unique set of advantages and disadvantages. The choice of method depends primarily on the demands imposed by the nature of the research being performed. We briefly describe the 2 most commonly used methods.

Mechanically scanned array

An MSA is essentially the result of directly automating the process performed when manually tracking. A single antenna or phased array is rotated automatically while signal strength is logged. After panning through the active sector in which the transmitter is assumed to be located, the variation of signal strengths is processed, and a bearing is calculated. The processing of signal strength variations becomes quite complex as it optimizes bearing accuracy.

A system of this type typically has high gain and hence long range. It has accuracy better than 5 degrees, but the accuracy degrades with an increase in speed of operation or a decrease in number of samples of signal strength collected on a given pan. Conversely, slow operation can contribute to a loss of accuracy due to transmitter movement. For example, a system may specify a bearing accuracy of 5 degrees if rotated through a given sector for 1 min. However, during this 1-min period, the animal being studied may travel 10 m. Therefore, at 100-m range from the receiving site there may be up to a 9-m uncertainty in position due to the 5-degree error, compounded with the 10-m uncertainty from animal movement. Experimentation at a given facility would determine particular site limitations. Parameters such as acquisition speed and position accuracy are compromised to optimize the system.

Switched yagi array

The SYA consists of an array of yagi antennas aimed in separate directions, fixed to the same mast, and coupled to a receiver through a computer-controlled electronic switch. The receiver cycles through the antennas in search of the maximum signal strength during a single pulse. Resolution can be increased by interpolating between antennas. This system has the advantages of reduced power consumption and increased acquisition rate over a mechanically scanned system and increased gain over a Doppler system.

Summary of system capabilities

Contemporary automatic systems, housed in 1 portable, ruggedized unit, are capable of using intelligent control algorithms to detect, measure, and record
  * absolute signal power,
  * cumulative presence and absence within defined zones (cells),
  * activity patterns,
  * nesting behavior,
  * approximate range,
  * time and date of arrival, and
  * temperature, mortality, and other sensor information.

Further, these systems' capabilities include functions that scan up to 8 antennas (cells or zones), either independently or in various combinations. The multiple antenna scanning procedure provides cellular positioning of the animal. System algorithms and adaptive software routines can, in real time, directly influence data collection methodologies and provide statistical overviews (databases) along with temporal and spatial information. Indeed, algorithms exist that provide the user complete strategic control over scanning methods and data collection. The use of master-and-slave antenna arrangements can be used to reduce databases further and provide a means of code collision separation in areas of dense transmitter populations.

In addition, available systems
  * utilize noise reduction and suppression routines,
  * recognize and detect various codes,
  * are field proven for harsh environment operation,
  * can be controlled remotely,
  * are able to automate direction finding,
  * run numerous application programs, and

- contain large memory for data storage.

The future—advances in signal processing

Commercial systems have recently taken a large step forward in technological development with the introduction of systems incorporating digital signal-processing capabilities. These breakthroughs vastly increase the capabilities of automatic systems and provide a bright light for future developments. The potential for these systems was examined and recognized over a decade ago (Kolz and Castles 1983), but not considered economical at that time.

New developments in computer technologies have significantly increased the processing power while decreasing the size and cost. Current capabilities include multi-channel monitoring of 24 channels, 7 antennas, and 150 codes simultaneously. Future developments in spread- spectrum signal technologies, providing code division multiple access and correlation processing, along with improved Fast Fourier analysis, will provide an order of magnitude improvement in range within the next 5 y.

References

Kolz AL, Castles MP. 1983. The development of correlation receivers for wildlife tracking. In: Proceedings, Fourth International Wildlife Biotelemetry Conference, Halifax, Nova Scotia, Canada. p 112–134.

Kolz AL, Johnson RE. 1981. The human hearing response to pulsed-audio tones: implications for wildlife telemetry design. In: Long FM, editor. Proceedings, Third International Conference on Wildlife Biotelemetry. Laramie: University of Wyoming. p 27–34.

Index

A

abbreviations and acronyms, xxii–xxiii
Accipiter cooperii, 69
acetylcholinesterase, 23, 72, 77, 80
acorn woodpecker, 62
activity recorders, xiii, 12, 43, 45
aerial application of pesticides, 59, 61
aerial photography, 34, 37
aerial surveillance, 39, 51, 98–99, 146
Agelaius phoeniceus, 55, 56, 57
Agricultural Stabilization Conservation
    Service (ASCS), 34
American kestrel. *See Falco sparverius*
American Radio Relay League, 133
American robin. *See Turdus migratorius*
analysis of deviance (ANODEV), 129
analysis of variance (ANOVA), 95, 96
Analysis of Wildlife Radio-Tracking Data,
    18
Anas platyrhynchos, 17, 25, 55, 56
Antenna Handbook, 133
antenna-height gain factor, 158–160, 184
antennas
    in avian telemetry, 34, 35–36, 37
    core, 136
    dipole, 138–139, 144, 148
    directional, 138
    dual yagi, 51, 61
    H-adcock, 138, 140–141, 148
    hand-held, 45, 51, 70, 146
    length of, 137–138, 142, 143
    loop, 136, 139
    mounts for, 146
    multielement, 139–140
    omnidirectional, 144
    receiving, 138–149

testing of field accuracy, 147–149
theory and practice, 133–149, 153–162
transmitting, 136–138
types of, 133
whip, 136, 137
yagi, 37, 39, 40, 70, 138, 140, 141–143, 145,
    185–186
    mounts for, 146
    multielement, 139–140
    omnidirectional, 144
    receiving, 138–149
    testing of field accuracy, 147–149
    theory and practice, 133–149, 153–162
    transmitting, 136–138
    types of, 133
    whip, 136, 137
    yagi, 37, 39, 40, 70, 138, 140, 141–143,
        145, 185–186
antenna tower systems, 44–45, 181
anti-cholinesterase compounds, 21, 22, 23,
    27
"application" of pesticide, regulatory
    definition, 11
Aquila chrysaetos, effects of Compound
    1080, 62
ARGOS platform transmitter terminal
    units, 163–172, 181, 187, 188
ash-throated flycatcher, 62
attenuation of signal, 135, 160–161. *See
    also* Signal loss
autocorrelation, 46
automatic detection finding (ADF), 187,
    188
automatic positioning systems (APS),
    187–188

voltage standing wave ratio (VSWR), 143,
    144–145
Vulpes fulva
  effects of Compound 1080, 62
  effects of strychnine, 57

W

weather factors, 162, 187
western meadowlark, 55, 56
white-breasted nuthatch, 62
Wildlife International, Ltd., 116
Wildlife Materials, Inc., 86, 89
Wildlife Radio Tagging, 18
Wilk's lambda, 177
World Wildlife Fund, xii

X

Xanthocephalus xanthocephalus, 55, 56

Y

yagi antennas. *See* Antennas, yagi
yellow-headed blackbird, 55, 56
yellow pine chipmunk, 58, 64

Z

Zenaida macroura
  effects of Compound 1080, 61
  effects of strychnine, 52, 53, 54
  effects of zinc phosphide, 60
zinc phosphide, 49, 50
  hazards to wildlife from baiting in
      orchards, 59–60, 63–64
  LD50, 59
z-test, 95